Crop Protection Strategies
for Subsistence Farmers

Studies in Insect Biology
Michael D. Breed, Series Editor

Crop Protection Strategies for Subsistence Farmers, edited by Miguel A. Altieri

Crop Protection Strategies for Subsistence Farmers

EDITED BY
Miguel A. Altieri

Routledge
Taylor & Francis Group

LONDON AND NEW YORK

First published 1993 by Westview Press, Inc.

Published 2018 by Routledge
52 Vanderbilt Avenue, New York, NY 10017
2 Park Square, Milton Park, Abingdon, Oxon OX14 4RN

Routledge is an imprint of the Taylor & Francis Group, an informa business

Library of Congress Cataloging-in-Publication Data
Crop protection strategies for subsistence farmers / [edited by]
   Miguel A. Altieri.
      p.   cm. — (Westview studies in insect biology)
   Includes bibliographical references and index.
   ISBN 0-8133-8635-7
   1. Plants, Protection of—Developing countries.   2. Agricultural
pests—Control—Developing countries.   3. Alternative agriculture—
Developing countries.   4. Farms, Small—Developing countries.
I. Altieri, Miguel A.   II. Series.
SB950.3.D44C76   1993
632'.9'091724—dc20   93-14990
                                                                                          CIP

British Library Cataloguing in Publication Data
A CIP catalogue record for this book is available from the British Library.

   ISBN 13: 978-0-367-01103-1 (hbk)
   ISBN 13: 978-0-367-16090-6 (pbk)

Contents

Preface

It is estimated that there are about 500 million small farmers throughout the developing world. In a constant struggle to survive, resource-poor farmers have developed diverse and complex cropping systems and management technologies that, in most instances, are well adapted to the rainfed and risk-prone environments in which they exist.

Promotion of chemical inputs, new agroexport and market pressures, and new government policies are creating rapid changes in the economic, environmental, and cultural context of small-scale agriculture. Thus, small farmers face new challenges, many of them of an unprecedented nature.

Conventional scientific research and extension programs have emphasized high-input approaches geared to modernizing small farm agriculture and, often, technological packages developed at experiment stations or transferred from industrialized countries that do not fit the conditions or needs of these farmers. In many cases, when adopted, these technologies cause environmental problems.

The field of integrated pest management has also been affected by these problems. In recent years the acute environmental and social costs associated with high-input crop protection technology have become increasingly obvious. Researchers and development specialists are desperately seeking new approaches to crop production and protection that will help alleviate rural poverty and halt environmental degradation. The articles in this book document a few efforts toward sustainable development for small farmers in Asia, Africa, and Latin America. It is interesting to note that most of these efforts, although conducted in very different socioeconomic as well as environmental milieus, emphasize farmers' participation, on-farm research, biological control techniques, and maintenance and enhancement of biodiversity at the farm level. These aspects seem to be crucial for the development of sustainable agriculture and rural development in the Third World.

Miguel A. Altieri

1

Designing and Improving Pest Management Systems for Subsistence Farmers

Miguel A. Altieri

Introduction

Traditional and/or peasant agriculture is a prominent rural activity in most parts of the developing world. For example, in Latin America there are about nine million peasant production units located mostly in marginal environments and exhibiting low productivity (Ortega 1986). Nevertheless, their contribution to regional food security is crucial since they produce most of the maize, beans, potatoes and other staple foods. This small farm sector has been bypassed by agricultural modernization, mainly because new technologies were not made available to small farmers on favorable terms and hence often they were not suited to the agroecological and socioeconomic conditions (de Janvry 1981). Pest management innovations are no exception (Altieri 1984 and 1985).

Critics of top-down rural development programs charge past IPM programs with a lack of understanding and appreciation of the agroecological, cultural and socio-economic milieu they operated in, exclusion of the small farmer as both collaborator and beneficiary, and inept promotion of inappropriate technology (Matteson et al. 1984). As a result the development and extension of improved and adaptable IPM technology for small farmers in developing countries is being re-examined. Among the various efforts to devise better crop protection methods it is possible to distinguish three main approaches:

1. Rescuing, understanding and applying traditional farming knowledge to solving agricultural and pest problems
2. Involving farmers in the design, conduction and evaluation of technologies through participatory research methods

3. Use of agroecological principles and techniques to design agroecosystems that enhance natural and biological control processes

The purpose of this chapter is to offer a conceptual foundation based on
the above three approaches for examining and optimizing small farming
systems in terms of crop production and protection concerns.

Ethnoecology and the Improvement of IPM
in Small Farming Systems

Interest in how traditional peasant farmers in developing countries perceive and modify their environment has recently increased among academic
institutions. Researchers from a variety of disciplines (i.e., anthropology,
human ecology, entomology, agronomy, soils, agroecology, ethnobiology,
etc.) are engaged in describing farmers' rationale, strategies to minimize
risk, local resource use, cropping systems design and management, folk
taxonomies, etc. Most of these studies have documented that in most areas
where cultural traditions and social organization have not been drastically
changed, farmers are excellent preceptors of their environment and make
successful management decisions designed to overcome production constraints (Brokensha et al. 1980, Klee 1980).

Evaluations on how traditional farmers perceive pest problems and on
the various indigenous control methods employed are few. The scattered
information is mostly observational/anecdotal and does not provide quantitative details about the effects of various cultural control practices on pest
dynamics or about the ecological mechanisms involved in the regulation of
specific pests. Nevertheless by assembling most of the current literature on
the subject, it is possible to offer a synthesis of current knowledge along
with an ecological basis to develop a theory on indigenous methods of pest
control.

Knowledge About Pests and Folk Taxonomies

Classification of animals, especially insects and birds, is widespread
among farmers and indigenous groups (Bulmer 1965). In their survey of
pest control practices used by local farmers in the Philippines, Litsinger et
al. (1980) found that farmers had local names in separate dialects in each
location for most pests attacking rice, corn and grain legumes. Farmers
were not aware of some pests considered as problems by entomologists,
and consequently did not attempt control measures.

Insects and related arthropods have major roles as crop pests, causes of
disease, food, medicinals, and are important in myth and folklore. In many
regions, agricultural pests are tolerated because they also constitute agricultural products; that is, traditional agriculturalists may consume plants

and animals that would otherwise be considered pests. In Indonesia, a grasshopper pest in rice is trapped at night and eaten (with salt, sugar, and onions) or sold as bird food in the market. Ants, some of which may be major crop pests, are one of the most popular insect foods gathered in tropical regions (Brown and Marten 1986).

In his studies of Kabba farmers in Nigeria, Atteh (1984) not only found that farmers could identify the pests affecting their crops, but that also they could rank the pests according to the degree of damage they caused to crops. In addition, further research revealed that for each pest farmers had knowledge of:

1. The history of the pest, including dates when the pest was noticed, when it became a menace, peak periods of occurrence in the past, and type of damage done;
2. The biology of the pest, including the life cycle of the pest, its breeding behavior, and ecological and climatic conditions facilitating or discouraging increase in numbers;
3. The bionomics of the pest—the feeding preferences and the severity of damage done to plants attacked.

A good example of farmers' knowledge of the biology and bionomics of pests is the case of the variegated grasshopper, *Zonocerus variegatus,* in southern Nigeria. Richards (1985) found that local farmer knowledge was equivalent to that of his scientific team concerning the grasshoppers' food habits, life cycle, mortality factors, degree of damage to cassava, and the egg-laying behavior and egg-laying sites of the females.

Farmers were aware that, numerous as these insects are, they congregate under only a few shaded areas on the farm or in an area to lay eggs at a particular period, and that these eggs are kept in pods and inserted an inch or so below the soft ground surface.

Farmers had discovered on their own that the egg-laying sites can be marked and the egg pods dug up. Once exposed to the hot sun the eggs die. They had in fact tried this as a control measure. Farmers had also established a close relationship between the presence of a weed (*Eupatorium odoratum*) and the advance and severity of the pest (Page and Richards 1977).

In this particular case farmers' knowledge added facts to that of the researchers in regard to the dates, severity and geographical extent of some of the outbreaks, plus the fact that the grasshopper was eaten and sold and was of special importance to women, children, and poor people. Thus the final control recommendation by scientists, clearing the egg-laying sites from a block of farms, did not require most farmers to learn new concepts, and for some the practice was nothing new (Richards 1985).

Indigenous Pest Control Methods

Traditional farmers rely on a variety of management practices to deal with agricultural pest problems. Two main strategies can be distinguished. One is the use of direct, non-chemical pest control methods (i.e., cultural, mechanical, physical and biological practices). The second is reliance on built-in pest control mechanisms inherent to the biotic and structural diversity of complex farming systems commonly used by traditional farmers (Brown and Marten 1986). Farmers also use a variety of other management practices that, although targeted for other farm purposes, significantly impact pest dynamics.

Traditional farmers throughout the world use a series of mechanical, cultural and biological measures to control pests (Brown and Marten 1986, Atteh 1984, Litsinger et al. 1980, Altieri 1985). Table 1.1 summarizes the main strategies as well as specific practices. This ensemble of cultural practices can be grouped into three main strategies, depending on which element of the agroecosystem is manipulated:

Manipulation of Crops in Time Farmers often carefully manipulate the timing of planting and harvest and use crop rotations to avoid pests. These techniques obviously require considerable ecological knowledge of pest phenology. Although these techniques often have other agronomic benefits (e.g., improved soil fertility), the farmers sometimes explicitly mention that they are done to avoid pest damage. For example, in Uganda farmers utilize time of planting to avoid stem borers and aphids in cereals and peas respectively. Many farmers are aware that planting out of synchrony with neighboring fields can result in heavy pest pressure and therefore use a kind of "pest satiation" to avoid extensive damage (mungbeans in the Philippines, Litsinger et al. 1980; rice in Indonesia, Prasadja and Ruhendi 1981). In the central Andes, a potato fallow rotation is carefully observed, apparently to avoid buildup of certain insects and nematodes (Brush 1983). Perhaps the most common way in which farmers manipulate the temporal permanence of agroecosystems is through the traditional pattern of slash and burn or shifting cultivation.

Manipulation of Crops in Space Farmers often manipulate plot size, plot site location, density of crops and crop diversity to achieve several production purposes, although most are aware of the links between such practices and pest control.

- Overplanting: One of the most common methods of dealing with pests is planting at a higher density than one expects to harvest. This strategy is most effective in dealing with pests that attack the plant during early stages of growth. When infested plants are detected, they are carefully removed long before actual death so as to avoid contaminating healthy plants.

TABLE 1.1   Pest management strategies and specific practices used by traditional farmers throughout the developing world.

Strategy	Practices
Mechanical and physical	Scarecrows, sound devices
	Wrapping of fruits, pods
	Painting stems, trunks with lime or other materials
	Destroying ant nests
	Digging out eggs/larvae
	Hand picking
	Removal of infested plants
	Selective pruning
	Application of materials (ash smoke, salt, etc.)
	Burning vegetation
Cultural practices	Intercropping
	Overplanting or varying seeding rates
	Changing planting dates
	Crop rotation
	Timing of harvest
	Mixing crop varieties
	Selective weeding
	Use of resistant varieties
	Fertilizer management
	Water management

- Farm plot location: In Nigeria many farmers, linked by kinship ties, age grouping or friendship, locate their farm plots lying contiguous to each other but leaving room for the expansion of each farm in a particular direction. In accounting for this practice, farmers reported that all pests in the area will discover and concentrate on an isolated farm. Plots are therefore grouped together to spread pest risk among many farmers (Atteh 1984). Conversely, in tropical America, Brush (1983) reports that farmers deliberately use small isolated plots to avoid pests. In many areas farmers carefully use environmental indicators in site location. For example, in Belize, areas where "tiger bush" grows are avoided since they indicate probable future pest problems, especially weeds.
- Selective weeding: Studies conducted in traditional agroecosystems show that peasants deliberately leave weeds in association with crops,

by not completely clearing all weeds from their cropping systems. This "relaxed" weeding is usually seen by agriculturalists as the consequence of a lack of labor and low return for the extra work; however, a closer look at farmer attitudes toward weeds reveals that certain weeds are managed and even encouraged if they serve a useful purpose. In the lowland tropics of Tabasco, Mexico, there is a unique classification of noncrop plants according to use potential on one hand and effects on soil and crops on the other. According to this system farmers recognized 21 plants in their cornfields classified as *mal monte* (bad weeds), and 20 as *buen monte* (good weeds) that serve as food, medicines, ceremonial materials, teas, soil improvers, etc. (Chacon and Gliessman 1982). Similarly, the Tarahumara Indians in the Mexican Sierras depend on edible weed seedlings (*Amaranthus, Chenopodium* and *Brassica*) from April through July, a critical period before maize, bean, cucurbits, and chiles mature in the planted fields in August through October. Weeds also serve as alternate food supplies in seasons when the maize crops are destroyed by frequent hailstorms. In a sense the Tarahumara practice a double crop system of maize and weeds that allows for two harvests: one of weed seedlings or *quelites* early in the growing season (Bye 1981). Some of these practices have important insect pest control implications since many weed species play important roles in the biology of herbivorous insects and their natural enemies in agroecosystems. Certain weeds, for example, provide alternate food and/or shelter for natural enemies of insect pests during the crop season but more importantly during the off season when prey/hosts are unavailable (Altieri and Whitcomb 1979).

Manipulation of Crop Diversity   The practice of multicropping and/or mixed cropping is intricately related to traditional agriculture (Toledo et al. 1985). Although most farmers use intercropping because of labor and land shortages or other agronomic purposes, the practice has obvious pest control effects (Altieri and Letourneau 1982). Many farmers know this and use polycultures as a play-safe strategy to prevent build-up of specific pests to unacceptable levels, or to survive in cases of massive pest damage. For example, in Nigeria, farmers are aware of the severe damage done to the sole crop of cassava by the variegated grasshopper after all other crops have been harvested. To reduce this damage farmers deliberately replant maize and random clusters of sorghum on the cassava plot until harvest time (Atteh 1984).

In his surveys of traditional maize cropping systems in Tlaxcala, Trujillo (1987) found that certain crop associations would reduce populations of the pestiferous scarab beetle *Macrodactylus* sp., while others would increase them. In a survey of insect communities associated with maize grown associated with other annual crops and with trees or shrubs in Indonesia, it was found that pest damage and abundance of natural enemies varied considerably between fields. It was clear that pest dynamics varied significantly between systems depending on insect species, location and size of the field, vegetational composition within and around fields and cultural management.

Manipulation of Other Agroecosystem Components

In addition to manipulating crop spatial and temporal diversity, farmers also manipulate other cropping system components such as soil, microclimate, crop genetics and chemical environment to control pests.

Use of Resistant Varieties Through both conscious and unconscious selection, farmers have developed varieties that are resistant to pests. This is probably the most widely used and effective of all the traditional methods of pest control. Litsinger et al. (1980) found that 73% of the peasant farmers in the Philippines were aware of varietal resistance even if they had not consciously tried to manipulate it. There is evidence in traditional varieties for all the modes of resistance that modern plant breeders select for, including pubescence, toughness, early ripening, plant defense chemistry and vigor.

In Ecuador Evans (1988) found that infestations in ripening corn ears by Lepidoptera larvae were significantly higher in new varieties than in traditional ones, a factor that influenced the adoption of new varieties by small farmers.

Water Management Manipulation of water level in rice fields is a widely used practice for pest control (King 1927). Water management is also practiced in many other annual crops for the same purpose. For example, in Malaysia, control of cutworms and armyworms is effected by cutting off the tip of infested leaves in a number of annual crops, and raising the water level, which carries the larvae into the field ridges, where birds congregate to eat them.

Plowing and Cultivation Techniques Farmers frequently report that they deliberately manage the soil (sometimes using more and sometimes less cultivation) to destroy or avoid pest problems. In Peru, for example, peasants use "high tilling" of potatoes to protect the tubers from insect pests and diseases (Brush 1983).

In shifting cultivation, after clearing a piece of land farmers set it on fire after a week or two, reportedly to reduce weed and pest populations during the first year of cropping (Atteh 1984).

Use of Repellents and/or Attractants Farmers have been experimenting with various natural materials found in their immediate environment (especially in plants) for many centuries, and a remarkable number have some pesticidal properties. A recent world review of plant extracts used in traditional pest control found 664 plant species in 135 families (Secoy and Smith 1983). An entire volume has been published on one of these species, the neem tree (*Azadirachta indica*), whose leaves and fruits repel sucking and chewing insects.

Use of plants or plant parts either placed in the field or applied as herbal concoctions for pest inhibition is widespread. Litsinger et al. (1980) interviewed 108 rice farmers in the Philippines and found that 43% placed plant parts in the fields to attract or repel insects. In Aloburo, Ecuador, small farmers place castor leaves in recently planted corn fields to reduce populations of a nocturnal tenebrionid beetle. These beetles prefer castor leaves over corn, but when associated with castor leaves for 12 hours or more, beetles exhibit paralysis. In the field, the paralysis prevents beetles from hiding in the soil, which increased their mortality by direct exposure to the sun (Evans 1988). In southern Chile, peasants place branches of *Cestrum parqui* in potato fields to repel *Epicauta pilme* beetles (Altieri and Farrell 1984). Many times a plant is carefully grown near the household and its sole function is apparently to provide the raw material for preparing a pesticidal concoction. In Tanzania, farmers cultivate *Tephrosin* spp. on the borders of their maize fields. The leaves are crushed and the liquid is applied to control maize pests (Risch, unpubl. manuscript). In Tlaxcala, Mexico, farmers "sponsor" volunteer Lupinus plants within their corn fields, because those plants act as trap crops for *Macrodactylus* sp. (Altieri and Trujillo 1987).

Manipulation of Natural Enemies   Ducks and geese have been used for many years in Java and China to control insect populations in rice fields (Chiu and Chang 1982). Chickens also have been used to control leaf-feeding insects on crops within home gardens (Brown and Marten 1986). In China nests of predaceous ants (*Oecophylla smaragdina*) were collected and placed in citrus groves to control insect pests. This practice was one of the first documented cases of biological control.

Involving Farmers in IPM Research and Extension

Many researchers and extension specialists know that all too often developed IPM technology is not adopted or used by farmers. On the other hand there are a number of technologies that did not originate in research stations and that are rapidly passed from farmer to farmer (Ashby 1990). Many researchers and NGO technicians have taken advantage of these farmer-initiated activities to develop IPM technologies for small farmers. Two case studies are described here to illustrate the potential of farmer involvement in technology design, diffusion and evaluation (van Schoubroeck et al. 1990).

IPM *in Wetland Rice*

FAO initiated an IPM program for rice in South and South-east Asia, in collaboration with the agricultural extension services and research institutes of the countries concerned. The program developed for Sri Lanka is presented here as an example. The aim of the program is to train rice farmers in IPM methods and also to ensure that the acquired knowledge is applied on the farms. It also aims to create a general awareness amongst the farming population of the negative effects of chemical control. In Sri Lanka a multi-media campaign has been developed for this purpose, in which, for example, radio announcements are used to create resistance to persuasive radio advertisements for pesticides (Hansen 1987).

The IPM training program is embedded in the structure of the government's agricultural extension services. This service operates through the so-called Training and Visit System, in which the extension officer at village level, the KVS (Krush Vyaphthi Sevaka) has 36 contact farmers under his or her care. These contact farmers undertake group training once every two weeks and should also be visited at home by the KVS. The IPM technology has become a part of a bi-weekly training program, which covers all aspects of rice cultivation. Each contact farmer is responsible for passing the acquired information on to 24 farmers.

Some five KVSs are trained and guided by an Agricultural Instructor (AI) who, in turn, is trained by others, including an expert on crop protection (SMO—Subject Matter Officer). All SMOs are intensively trained in IPM through the FAO program, as are a selected group of AIs and the KVSs, thus maintaining direct contact with the people working at village level. One advantage of this extension system, when it works well, is that many farmers are reached quickly. However, the danger of superficiality in the transfer of information through so many levels is great. To eliminate this problem, the FAO organizes, in addition to the regular bi-weekly training programs for contact farmers, training programs lasting a full season. KVSs give specific training in IPM, to a fixed group of 20–25 farmers who have applied for this training, once a week during the growing season. A much greater amount of information can be passed on in these training courses.

During the years 1984–1986 the FAO project in Sri Lanka has extended to 19 of the total of 26 districts. In 1986, 17,000 farmers were trained in each season.

The training programs for IPM cover an extensive set of recommendations, the most important of which, as applicable to rice cultivation include:

- timely weed control
- adequate fertilization

- use of resistant varieties
- synchronous planting in adjoining fields
- protection and stimulation of natural enemies by avoiding use of toxic pesticides and by providing nectar-bearing plants.

It is expected that by using the techniques of IPM, as disseminated by the FAO, higher profits should be gained by rice farmers, either through savings on pesticides or through higher yields as a result of more effective pest management. The whole range of recommendations makes a great demand on the farmers as far as learning and use of new methods is concerned. However, many of the recommended practices are already used traditionally by the farmers, such as thorough cultivation of the soil. The farmers must become aware of the value of such practices in the context of crop protection.

The most important aspect of IPM, the weekly sampling of the crop, is especially time consuming for the majority of farmers. The IPM approach should, in theory, not cause any additional costs, but money must be available at the right time to buy the required inputs. This could prove to be a problem particularly for the major irrigated farmer. A further condition is that the required inputs, such as resistant varieties and selective pesticides, should be available locally and at the right time. This is often not the case for rainfed farmers, who live in the more distant, inaccessible mountain regions. Farmers who are difficult to reach are also visited less often by extension officers, and have greater difficulty in attending training sessions.

IPM of the Corn Earworm in the Andes

Since the early 1970s, the corn earworm *Heliothis zea* is the most significant maize pest in the mountainous part of Ancash in the northern Peruvian Andes (van Schoubroeck et al. 1990). The biggest problem in the control of corn earworm is the short time during which the eggs and larvae are on the surface of the cob and can thus be reached by a pesticide. Once the larvae have eaten their way inside the cobs there is no point in spraying with pesticides. The usual control method advised by the extension services consists of repeated spraying of the pistils at intervals of three days which results in an average of six sprayings per field with a total of about 16 kg per hectare of Sevin (85% carbaryl) or the organophosphorous compound Dipterex (80% trichlorphon). For rainfed maize farmers at zones higher than 2200 meters implementation of this spraying advice is barely possible. A large part of the maize production is intended for self-consumption and this makes expenditure on knapsack sprayer and pesticides very cost ineffective. Rainfed farms suffer the greatest damage from the corn earworm because where genetically heterogeneous local varieties are sown, the period during which the maize is susceptible is longer than with hybrid seeds which produce uniform plants developing at the same rate. Although the local varieties are in themselves not more susceptible to this pest, the chance of infestation is greater and the execution of a spraying program is more expensive. On the other hand wide adoption of hybrids can lead to higher costs due to higher fertilizer requirements. A potential danger is the displacement of local cultivars enhancing genetic erosion.

Early attempts to control the corn earworm included three strategies: the use of the parasitic wasp *Trichogramma*, intensive tillage to kill the pupae and the injection of a pesticide into the maize cobs. Research evaluations of the efforts showed that the biological control of corn earworm with *Trichogramma* is not effective in practice, as these wasps do not reproduce under field conditions. Killing the pupae by repeated tillage also had no effect as the adult moths move throughout the area, so there is a constant supply of new moths which fly in from elsewhere to lay their eggs. Moreover the *Heliothis* problem has increased in the last few years, because irrigated cultivation is now carried out throughout the whole year, promoting the build-up of the pest population through lack of a crop-free period.

The most practical way of limiting the damage caused by the corn earworm proved to be chemical control by means of injection. In practice the approach consists of injecting within eight days after the first eggs have been sighted on ten sampled plants. In this period there should be daily control on whether the number of eggs increases or not. If there are doubts about the extent of the infestation, fifty plants are sampled. This means that the tops of the cobs are carefully opened and checked to see whether the small larvae (of about 6 mm) have penetrated. If this is the case in more than ten cobs, then the action threshold has been reached and the infestation is sufficiently severe to warrant control. The injection method uses the carbamate Sevin mixed into a solution ten times as concentrated as for normal spraying. This solution is injected into every corn cob with a disposable veterinary syringe fitted with a thick needle. The dose of 1 cc per cob must be strictly followed. More than this amount leads to rotting through too much moisture inside the cobs. Less than 1 cc per cob or a more diluted solution does not kill enough larvae. The most important advantage of the injection method is that a knapsack sprayer is no longer necessary and that much less pesticide is used; an average of 2.5 kg carbamates per hectare, as compared to the 16 kg required by the method recommended by the agricultural extension services. In addition to this, the method causes less harm to the environment through more localized use of the pesticide and a reduced dosage. For years after these recommendations were first made, the injection method is used on an estimated 30% of the maize farms. This number is steadily increasing through the exchange of information between farmers. The agricultural extension service has since then also incorporated the method in its training programs (Hansen 1987).

Applying Agroecology to Pest Management in Small Farming Systems

In many LDCs there is a clear need to build new research and extension capabilities that translate into specific actions which actually improve the livelihood of resource-poor farmers. Efforts in Latin America during the last 10–15 years led by non-governmental organizations (NGOs) are providing some important lessons and guidelines. The approach of many NGOs has been to search for technologically unconventional systems of agricultural development that are based on local participation, resources and skills, as well as on modern agricultural science, and enhanced productivity while conserving the resource base (Altieri and Anderson 1986). This new agroecological development approach is more sensitive to the complexities of local agricultures, emphasizing properties of sustainability, food security, biological stability, resource conservation and equity, along with the goal of increased production.

In practical terms, the application of agroecological principles has translated into programs that emphasize (Altieri 1991a):

- Improving the production of basic foods, including the traditional food crops (Amaranthus, quinoa, lupine, etc.) and the conservation of native crop germplasm;
- Recovering and re-evaluating peasants' knowledge and technologies;
- Promoting the efficient use of local resources (land, labor, minor agricultural products, etc.);
- Increasing crop and animal diversity to minimize risks;
- Improving the natural resource base through water and soil conservation and regeneration practices;
- Reducing the use of external chemical inputs, testing and implementing organic farming and other low-input techniques.

Promoted agroecological techniques are culturally compatible since they do not question peasants' rationale, but actually build up on traditional farming knowledge and combine it with modern scientific science. Also, the techniques are ecologically sound since they do not attempt to radically modify or transform the peasant ecosystem, but rather optimize it and conserve the resource base. Costs of production are minimized by enhancing the use efficiency of locally available resources, including the biodiversity inherent to local farms (Altieri 1991a).

Using Appropriate Cropping System Design

Many NGOs have taken on the challenge of designing cropping systems adapted to marginal environments and that are able to sponsor their own soil fertility, crop protection and stable yields. The use of diversity is not foreign to traditional agriculturalists. By mimicking natural ecological processes, farmers have evolved complex "agroecosystems," the sustainability of which has stood the test of time. Moreover, unlike modern monocultures, traditional agroecosystems reflect the priorities of peasant farmers; they produce a varied diet, achieve a diversity of sources income, use locally available resources, minimize the risk to farmers from crop losses, protect against the incidence of pests and disease and make efficient use of available labor. Such multiple cropping methods are estimated to provide as much as 15–20 percent of the world's food supply. Throughout Latin America, farmers grow from 70–90 percent of their beans with maize, potatoes and other crops. Maize is intercropped on 60 percent of the region's maize-growing area. Much of the experimental evidence suggests that insect pests (particularly specialized herbivores) exhibit lower population densities in polycultures than in corresponding monocultures (Altieri 1991b). In monoculture systems, herbivores exhibit greater colonization rates, greater reproduction, less tenure time, less disruption of host finding and reduced mortality by natural enemies (Altieri 1990).

Maintaining Genetic Diversity

Traditional agroecosystems are also genetically diverse, containing numerous varieties of domesticated crop species as well as their wild relatives. In the Andes, farmers cultivate as many as 50 potato varieties in their fields. Maintaining genetic diversity appears to be of even greater importance as land becomes more marginal, and hence farming more risky. In Peru, for example, the number of potato varieties cultivated increases with the altitude of the land farmed. Genetic diversity confers at least partial resistance to diseases that are specific to particular strains of crops and allows farmers to exploit different micro-climates for a variety of nutritional and other uses.

Traditional farmers also know the potential value of weeds in controlling pests, chiefly by providing a habitat for beneficial insects and/or acting as trap crops or repellent plants. For example, in Colombia, grass-weeds (*Elseusina indica* and *Leptochloa filiformis*) are grown around small bean fields to repel *Empoasca kraemeri* leafhoppers, a serious bean pest. In Tlaxcala, Mexico, farmers encourage the growth of Lupinus plants within their corn fields, as they attract the scarab beetle *Macrodactylus* away from the corn plants (Altieri 1991).

Diversity is not only maintained within the area cultivated. Many peasants maintain uncultivated areas (such as forests, lakes, grasslands, streams and swamps) in or adjacent to their fields, thus providing valuable products including food, construction materials, medicines, organic fertilizers, fuels and religious items. Where shifting cultivation is practiced, the clearing of small plots from secondary forest vegetation also permits the easy migration of natural pest predators from surrounding forest (Matteson et al. 1984).

Promoting Sustainable Agriculture

A major strategy used by NGOs in promoting sustainable agriculture is to enhance and/or restore agricultural diversity in time and space through crop rotations, cover crops, intercropping, crop/livestock mixtures, etc. Different options to diversify cropping systems are available depending on whether the current monoculture systems to be modified are based on annual or perennial crops. Diversification can also take place outside of the farm, for example, in crop-field boundaries with windbreaks, shelterbelts, and living fences, which can improve habitat for wildlife and beneficial insects, provide sources of wood, organic matter, resources for pollinating bees, and, in addition, modify wind speed and the microclimate (Altieri and Letourneau 1982).

There are many alternative diversification strategies which exhibit beneficial effects on soil fertility, crop protection and crop yields. If one or more of these alternative technologies are used, the possibilities of complementing interactions between agroecosystem components are enhanced, resulting in one or more of the following effects:

- continuous vegetation cover for soil protection
- constant production of food, ensuring a varied diet and several marketing items
- closing of nutrient cycles and effective use of local resources
- soil and water conservation through mulching and wind protection
- enhanced biological pest control through diversification
- increased multiple use capacity of the landscape
- sustained crop production without use of environmentally degrading chemical inputs.

An example of the application of diversity principles to design sustainable systems adapted to the needs of small farmers is the effort of the Centro de Educacion y Tecnologia (CET) in Chile. In an attempt at helping resource-poor peasants to achieve year-round food self-sufficiency and rebuilding the productive capacities of their small land holdings, CET's approach has been to establish several 0.5 ha model farms where most of the food requirements for a family with scarce capital and land can be met. In this farm system, the critical factor in the efficient use of scarce resources is diversity (Altieri and Anderson 1986).

The CET farm is a diversified combination of crops, trees, and animals. In an attempt to maximize production efficiencies, the components are structured to minimize losses to the system and promote positive inter-actions. Thus, crops, animals, and other farm resources are managed to optimize production efficiency, nutrient and organic matter recycling, and crop protection. The principal components are:

- Vegetables: spinach, cabbage, tomatoes, lettuce, etc.
- Intercropped maize/beans/potatoes and peas/fava beans
- Cereals: wheat, oats, barley
- Forage crops: clover, alfalfa, ryegrass
- Fruit trees: grapes, oranges, peaches, apples, etc.
- Non-fruit trees: black locust, honey locust, willows, etc.
- Domestic animals: chickens, pigs, ducks, goats, bees, and one dairy cow.

The animal and plant components are chosen according to (a) the crop or animal nutritional contributions, (b) their adaptation to local agro-climatic conditions, (c) local peasant consumption patterns and (d) market opportunities. The design is also based on cropping patterns, crops, and management techniques practiced by local campesinos. In these systems, campesinos typically produce a great variety of crops and animals. It is not unusual to find as many as 5 or 10 tree crops, 10–15 annual crops, and 3–5 animal species on a single farm.

The physical layout of these farms varies depending on local conditions; however, most vegetables are produced in heavily composted raised beds located in the garden section, each of which can produce up to 83 kgs of fresh vegetables per month. The rest of the vegetables, cereals, legumes, and forage plants are produced in a six-year rotational system (Figure 1.1). This rotation was designed to provide the maximum variety of basic crops in six plots, taking advantage of the soil-restoring properties of rotations. Each plot received the following treatments during the six-year period:

Year 1	Summer: corn, beans, and potato Winter: peas and fava beans
Year 2	Summer: tomato, onion, squash Winter: supplementary pasture (oats, clover, ryegrass)
Year 3	Summer: soybean, peanuts, or sunflower Winter: wheat companion-planted with pasture
Year 4-6	Permanent pasture: clover, alfalfa, ryegrass

In each plot, crops are grown in several temporal and spatial designs, such as strip-cropping, intercropping, mixed-cropping, cover crops, and

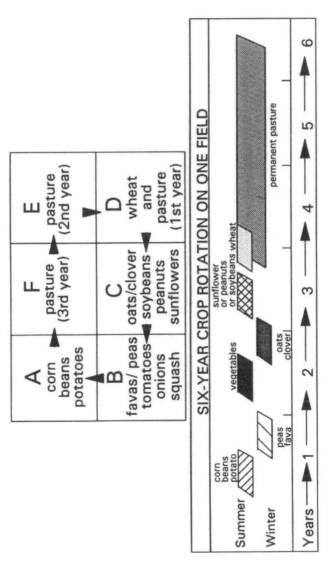

FIGURE 1.1   Spatial and temporal design of a complex six-year rotational cropping system aimed at sustained yield and enhanced crop protection in central Chile.

living mulches, thus, optimizing the use of limited resources and enhancing the self-sustaining and resource-conserving attributes of the system. An important consideration in the rotational design was the stability of the cropping system in terms of soil fertility maintenance and pest regulation. It is well accepted that a rotation of grains and leguminous forage crops provides significant additional inputs of nitrogen and much higher yields of the subsequent crop of grain obtained with continuous grain mono-cropping. The output of grain depends on the efficiency of the legumes in supplying nitrogen. Studies in Chile, and elsewhere, have shown that legumes such as sweet clover, alfalfa, and hairy vetch can produce between 2.3–10 tons per ha of dry matter and fix between 76–367 kg of nitrogen per ha. This is sufficient to meet most of the nitrogen requirements of agronomic and vegetable crops.

The rotational scheme provides nearly continuous plant cover that aids in the control of annual weeds. Incorporating legume cover crops in annual crops, such as corn, cabbage, and tomato, through overseeding and sod-based rotations has reduced weeds significantly. In addition, these systems reduce erosion and conserve moisture and, therefore, offer great potential for hillside farmers.

The crop rotations scheme also had a profound impact on insect pest populations. For example, the corn rootworm (*Diabrotica* spp.) consistently reached higher levels in a continuous corn monoculture than in corn fields following soybean, clover, alfalfa, or other crops. The presence of alfalfa in the rotational scheme enhanced the abundance and diversity of insect predators and parasites on the farm. Strip cutting of alfalfa forced movement of predators from alfalfa to other crops. Cutting and spreading alfalfa hay, containing high numbers of beneficial insects throughout the farm, also increased natural enemy populations. Cereal residues used as straw mulches in the succeeding crops significantly reduced white fly populations, the principal vector of several viruses, by affecting their attraction to host crops. Spiders, ground beetles and other predators also were enhanced by the alternative habitat provided by the mulch.

CET personnel have closely monitored the performance of this integrated farming system. Throughout the years, soil fertility has improved (P_2O_5 levels, which were initially limiting, increased 5–15 ppm) and no serious pest or disease problems have been noticed. The fruit trees in the garden orchard and those surrounding the rotational plots produce about 843 kgs of fruit/year (grapes, quince, pears, plums, etc.). Forage production reaches about 18 tons/0.21 ha per year, milk production averages 3,200 lts per year, and egg production reaches a level of 2.531 units. A nutritional analysis of the system based on the production of the various plant and animal components (milk, eggs, meat, fruits, vegetables, honey, etc.) shows that the system produces a 250% surplus of protein, 80% and 550% surplus of vitamin A and C respectively and a 330% surplus of calcium. A household economic analysis indicates that given a list of preferences, the balance between selling surpluses and buying preferred items is a net income of U.S. $790. If all the farm output is sold at wholesale prices, the family could generate a net income of U.S. $1635, equivalent to a monthly income of U.S. $136, 1.5 times greater than the monthly legal minimum wage in Chile.

Conclusions

It has become clear that some rural development programs are unsuccessful because the introduction of high-input technology often proves inappropriate. Recommendations, aimed at improving the short-term economic performance of the cropping systems, invariably furthers the use of pesticides and other capital intensive technology, often beyond the reach of resource-poor farmers. It is important to realize that most peasant farmers do not seek to maximize yields through the use of external inputs but rather to achieve long-term yield stability through diversity. In this regard, traditional farming systems exemplify efficiency and the careful management of soil, water, nutrients and biological resources. Strengthening such systems—through village-based initiatives that actively involve local peasants—is the key to successful grassroots rural development programs.

In this regard, the ensemble of traditional crop protection practices used by small farmers represents a rich resource for modern workers seeking to create pest management systems that are well adapted to the agroecological and socioeconomic circumstances of peasants. Peasants use a diversity of techniques, many of which fit local conditions well. The techniques tend to be knowledge-intensive rather than commodity-intensive. That is, their effectiveness is dependent on detailed knowledge of the agricultural environment rather than the use of expensive inputs. Also the techniques seem to have little negative environmental impact, which is a significant advantage over "modern" pest control strategies, which rely on expensive and toxic chemicals that often lose effect over a short period of time. Clearly, not all traditional crop protection components are effective or applicable; therefore modifications and adaptations may be necessary, but the foundation of such modifications must be based on peasants' rationale and indigenous knowledge. A reevaluation of traditional crop management, which has hitherto assured stable productivity on a long-term basis, will be necessary to develop more sustained and low input IPM practices and to reverse some of the "ecological crisis" triggered by modern changes (i.e., large-scale monocultures, promotion of high-yielding varieties, use of pesticides, etc.).

More emphasis on on-farm research which involves small farmers in managing experimental IPM technology is crucial to introduce farmers' criteria and rationale into the assessment of technology. Farmers' involvement provide researchers with direct insights into farmers' priorities and how farmers choose among several technologies. The final goal, however, is that farmers become social actors capable of determining the direction of their "own development."

Bibliography

Altieri, M. A. 1984. Pest management technologies for peasants: A farming systems approach. *Crop Protection* 3: 87–94.

————. 1985. Developing pest management strategies for small farmers based on traditional knowledge. *Bulletin of the Institute for Development Anthropology* 3: 13–18.

————. 1990. "The ecology and management of insect pests in traditional agroecosystems," in *Ethnobiology: Implications and Applications*. D. A. Posey and W. L. Overol (eds.). Proc. First. Int. Congress of Ethnobiology, Belem, Brasil. pp. 131–142.

————. 1991a. Traditional farming in Latin America. *The Ecologist* 21: 93–96.

————. 1991b. "Ecology of tropical herbivores in polycultural agroecosystems," in *Plant-Animal Interactions: Evolutionary Ecology in Tropical and Temperate Regions*. P. W. Price, T. M. Lewisohn, G. Wilson Fernandes and W. W. Benson (eds.). John Wiley and Sons, New York.

Altieri, M. A. and K. M. Anderson. 1986. An ecological basis for the development of alternative agricultural systems for small farmers in the Third World. *American Journal of Alternative Agriculture* 1: 30–38.

Altieri, M. A. and J. Farrell. 1984. Traditional farming systems in south-central Chile with emphasis on agroforestry. *Agroforestry Systems* 2: 33–18.

Altieri, M. A. and D. K. Letourneau. 1982. Vegetation management and biological control in agroecosystems. *Crop Protection* 1: 405–430.

Altieri, M. A. and J. Trujillo. 1987. The agroecology of corn production in Tlaxcala, Mexico. *Human Ecology* 15: 189–220.

Altieri, M. A. and W. H. Whitcomb. 1979. The potential use of weeds in the manipulation of beneficial insects. *HortScience* 4: 401–405.

Ashby, J. A. 1990. "Evaluating technology with farmers." Centro Internacional de Agricultura Tropical, IPRA Projects. Cali, Colombia. 95 p.

Atteh, O. D. 1984. Nigerian farmers' perception of pests and pesticides. *Insect Science and Application* 5: 213–220.

Brokensha, D., D. M. Warren & O. Werner (eds.). 1980. *Indigenous Knowledge Systems and Development*. University of America Press, Washington.

Brown, B. J. and G. G. Marten. 1986. "The ecology of traditional pest management in southeast Asia." East-West Center, Honolulu, Hawaii: Working Paper.

Brush, S. B. 1983. Traditional agricultural strategies in the hill lands of Tropical America. *Culture and Agriculture Newsletter* 18: 9–16.

Bulmer, R. 1965. Review of Navajo Indian ethnoentomology. *American Anthropology* 67: 1564–1566.

Bye, R. A. 1981. Quelites—ethnoecology of edible greens—past, present and future. *Journal of Ethnobiology* 1: 109–114.

Chacon, J. C. and S. R. Gliessman. 1982. Use of the "non-weed" concept in traditional tropical agroecosystems of south-eastern Mexico. *Agroecosystems* 8: 1–11.

Chiu, W. F. and Y. H. Chang. 1982. Advances of science of plant protection in the People's Republic of China. *Annual Review of Phytopathology* 20: 71–92.

de Janvry, A. 1981. *The Agrarian Question and Reformism in Latin America.* The John Hopkins Univ. Press, Baltimore. 311 p.

Evans, D. A. 1988. "Insect pest problems and control strategies appropriate to small–scale corn farmers in Ecuador." Ph.D. Dissertation. Univ. of Calif., Davis. 121 p.

Hansen, M. 1987. *Escape From the Pesticide Treadmill: Alternatives to Pesticides in Developing Countries.* Inst. Consumer Policy Res., Mt. Vernon, N.Y.

King, F. H. 1927. *Farmers of Forty Centuries.* Cape, London.

Klee, G. A. 1980. *World Systems of Traditional Resource Management.* John Wiley and Sons, N.Y.

Litsinger, J. A., E. C. Price and R. T. Herrera. 1980. Small farmer pest control practices for rainfed rice, corn and grain legumes in three Philippine provinces. *Philippine Entomology* 4: 65–86.

Matteson, P. C., M. A. Altieri and W. C. Gagne. 1984. Modification of small farmer practices for better pest management. *Annual Review of Entomology* 29: 383–402.

Ortega, E. 1986. "Peasant agriculture in Latin America and the Caribbean." Joint ECLAC/FAO Agriculture Division, Santiago, Chile.

Page, W. and R. Richards. 1977. Agricultural pest control by community action: the case of the variegated grasshopper in southern Nigeria *African Environment* 2: 127–141.

Prasadja, I. and Ruhendi. 1981. "Cropping systems entomology research in Indonesia." Proc. 11th Asians' Cropping Systems Work Group (1981). Puncak, Bogor.

Richards, P. 1985. *Indigenous Agricultural Revolution.* Westview Press, Boulder, Colorado.

Secoy, D. M. and A. E. Smith. 1983. Use of plants in control of agricultural and domestic pests. *Economic Botany* 37: 28–57.

Toledo, V. M., M. J. Carabias, C. Mapes and C. Toledo. 1985. *Ecologia y Autosuficiencia Alimentaria.* Siglo Veintiuno, Editorial, Mexico.

Trujillo, J. 1987. "The agroecology of maize production in Tlaxcala, Mexico: cropping system effects on arthropod communities." Ph.D. Dissertation. Univ. Calif., Berkeley. 146p.

van Schoubroeck, F.H.J., M. Herens, W. DeLeouw, J. M. Louwen and T. Overtown. 1990. *Managing Pests and Pesticides in Small Scale Agriculture.* Centre for Development Work, The Netherlands.

2

Steps Toward an Alternate and Safe Pest Management System for Small Farmers in the Philippines

Kenneth T. MacKay, Candida B. Adalla, and Agnes Rola

Introduction

Crop losses from pests (diseases, insects and weeds) in the developing countries can be high. Estimated annual losses due to pests throughout Southeast Asia vary from 10% to 30% depending on crop and environment (Teng and Heong 1988). These losses can seriously affect household income and nutrition.

Effective pest management is essential. However, conditions in developing countries are often much different from those in the developed countries so technology, pesticides, and approaches suitable for the developed countries often do not work.

The intensification of agriculture associated with the "green revolution" while increasing cropping intensity and yields has also created ideal conditions for insect pests. In irrigated areas where seasonal planting of rice has changed to asynchronous planting, there has been an increase in pest populations particularly of specialized pests (Loevinsohn and Litsinger 1989). A similar situation exists in vegetable growing areas where vegetables are planted year round. The year round availability of host plants plus the high temperatures and humidity of the humid tropics offer ideal conditions for pest multiplication. The diamondback moth (*Plutella xylostella*) for example has up to 28 generations per year in Malaysia (Ho 1965) as compared to 2 to 3 in temperate environments.

The agricultural landscape in developing countries is much more complex and diverse in both time and space than the monocultures typical of developed country agriculture. Land holdings are usually small, frequently ranging from 0.5 to 4 ha. Cropping systems in developing countries are very complex. While monoculture may dominate some agroecosystems during a portion of the cropping year, multiple cropping and intercropping are also common. In addition, one farm family may be growing different crops in different parcels of land at the same time. Some of the crops may be grown for home use (subsistence crops) while others for the market (cash crops). Pest management often differs between subsistence and cash crops. Pest management technology packages and loans are available for the cash crops but not for subsistence crops. In addition, because of the risk associated with the cash crops, farmers are unwilling to test new approaches on these crops.

In developing countries, typically 50 to 75% of the population are directly engaged in agriculture. Many farmers are poor and have no cash for purchase of inputs and little access to formal credit. Informal credit may be available from "middlemen" but interest rates are high. It is only in the major cash crops (vegetables, rice, wheat, etc.) where government-supported credit schemes are available for the purchase of inputs.

Farmers also have very little equipment. Sophisticated sprayers are not affordable or maintainable by small farmers. The backpack sprayer is normally used to apply chemical pesticides, however, these sprayers are often ill-maintained and leak resulting in high toxic exposure to the operator. Protective clothing is normally not available or worn.

Farmers often have little formal education, may be illiterate, and in some countries speak a language or dialect different from the national language. They, however, have considerable indigenous traditional knowledge about agricultural practices which is usually ignored by agricultural researchers and extension agents.

Research and extension efforts directed towards pest management are limited. There is no international institute or board which is devoted to developing appropriate pest management techniques for developing countries and very often methods and technologies from developed countries are transferred with little "in country" testing. Extension agents have very little knowledge of pest management and the ratio to farmers is low.

In this chapter we will first examine the problems of the most commonly used pest management technique, the use of synthetic pesticides, then explore some alternative approaches and examine the difficulties of applying these alternatives in the Philippines, and finally suggest some future directions that will lead to a safer and more sustainable agricultural approach.

Synthetic Pesticides

Synthetic pesticides (insecticides, herbicides, fungicides, etc.) are now used widely in developing countries. Their use is expanding and is expected to double in the next 10 years. They are used intensively in plantation and large scale farming activities, and also by small scale farmers for cash crops of grain, cotton, vegetables and fruits for both local and export markets. It is in the "green revolution" areas where grain (rice, maize, and wheat) production has increased dramatically that technical assistance and credit have been available for pest management, and it is here that pesticide use has increased substantially. Conversely, subsistence crops receive little or no pesticides.

The health risk to small farmers from use of synthetic pesticides is much greater in developing countries as compared to developed countries. The backpack sprayer results in much greater operator exposure than the more sophisticated and larger scale equipment used in developed countries. The sprayers often leak or are poorly maintained. Loevinsohn (1989) shows with data from a Malaysian rice growing area that 58% of the sprayers were corroded and 48% had the tank dented or cracked with 25% leaking from the tank valve or hose. Suitable protective clothing is not available or can't be worn because of the heat and humidity. Locally available clothing such as handkerchiefs or shawls as respirators and cotton gloves when worn, quickly become saturated and increase exposure still further.

Pesticide residues may be determined in export crops but there is no routine analysis of food intended for local consumption. Vegetables and fruits receive the highest pesticide doses and may contain unacceptably high residues. However, in South East Asia most of the "officially" published papers indicate pesticide residues are normally below FAO acceptable levels (Hashim and Yeoh 1988, Ramos-Ocampo et al. 1988, Soekardi 1988, Tayaputch 1988). Some less official reports indicate that vegetables frequently contain unacceptable levels of pesticides. Some farmers spray right up to harvest and even apply fungicides during transport to market. Rola (1989) documents vegetable farmers in the Philippines increasing spray applications as the crop approached harvest and dipping freshly picked vegetables in formalin to maintain consumer appeal. She also shows that banned or restricted organochlorines such as DDT, dieldrin, aldrin, etc. were present in most vegetables, sometimes in excess of allowable limits. Similar results have been found in Thailand where organochlorine residues (particularly DDT, Dieldrin and Heptachlor) above FAO acceptable limits were found in peanut seeds purchased in the markets (Wanleelag and Tau-Thong, 1986 and 1987; Wanleelag et al, 1988).

Pesticides required in small quantities are often repackaged in pop bottles and paper bags with no labels, stored near food and often reused as food containers. Even when labels are applied they may not be in the local language or farmers can't read them. Pesticides are the most common agent of suicide and accidental poisoning. In Sri Lanka, two thirds of the 25,000 annual cases of poisoning are due to pesticides.

Public health risks of pesticides use in the developed countries are greatly under reported. Previous estimates of 10,000 pesticide related deaths/year have been revised (Jeyaratnam et al., 1987; Loevinsohn, 1987) and indicate much more serious public health consequences of pesticide use than previously accepted. Loevinsohn (1987) reports a 27% increase in deaths among the most at-risk population in Central Luzon, Philippines after the introduction of pesticides.

Loevinsohn (1989) argues that many pesticides considered safe under developed country conditions are unsafe under tropical developing country conditions. The importance of these conclusions are that no matter how much there are improvements in applicator safety and education of farmers, certain chemicals are still too dangerous to use. The above conclusions relate primarily to classes of chemicals (organophosphates) which are considered to be more toxic to humans than more recently developed insecticides. However, recent results from China indicate that pyrethroids which are considered to have lower human toxicity also can have considerable effects on applicators under tropical conditions (Charbonneau, 1989).

There is little hard evidence of the wider environmental effects of pesticides on non-target organisms in developing countries. In Southeast Asia there is anecdotal evidence for loss of diversity particularly lower numbers of beneficial insects, and in rice growing areas considerable reduction in frogs and fish in rice paddies and irrigation water in areas of pesticide use.

The development of insect resistance to chemical pesticides is also a major problem. There may be as many as 1400 insect pests resistant to synthetic insecticides. *Plutella xylostella*, the diamond back moth (DBM) is one of the most resistant in S. E. Asia. It is resistant to all classes of synthetic insecticides, shows cross resistance to several, (Rushtapakornchai and Vattanatangum, 1986; Cheng, 1986) and may be resistant to the new class of insect growth regulators and *Bacillus thuringiensis* (Bt.). Resistance to insecticides is compounded by the farmers practice of mixing different insecticides and using application rates much higher than recommended. The recent outbreaks of the brown planthopper (BPH) (*Nilaparvata lugens*) on rice are clearly chemical-induced (Kenmore et. al. 1984, Heinrich and Mochida 1984, Reissig et. al. 1982, Litsinger 1990). This breakdown in resistance was caused by frequent application of insecticides especially methyl parathion, monocrotophos, azinphosmethyl and cypermethrin. These chemicals, when used as foliar sprays reduce the natural enemies and induce BPH outbreaks (Heinrich et al. 1982, Kenmore et. al. 1984). However, even with this knowledge among technical scientists, governments continued to recommend use of calendar application of these insecticides.

The conclusion from the above is that synthetic chemical pesticides as currently used are not the answer for pest management under tropical conditions. In addition the farmers who use and pay for pesticides are also the direct recipients of the external factors of chemical pesticides. They are the direct sufferers from the health and environmental side effects of chemical pesticide use.

Solutions

Solutions to safe pest management are desperately needed. Various alternative pest management approaches have been suggested and tried to decrease the reliance on chemical pesticides. Actions are needed at the international, national, agroecosystem and farm levels in order to decrease the use of synthetic pesticides and increase farmer, consumer and environmental health. These actions are:

1. Policies by national government which are supportive of the development of alternatives to chemical control;
2. Advocacy by non-governmental organizations (NGO's) at the national and international level to effectively lobby and raise issues often ignored by government bodies, and to support national policy;
3. Agroecosystem level changes for the management of fields, crops and water;
4. Appropriate pest management tools; and
5. Improved pest management decision making at the farm level.

Policy

In order for the development of an improved pest management system there is a need for appropriate national policies. These policies must include the appropriate pricing of pesticides to remove the distortion that subsidies produce (Repetto 1985); effective regulation of pesticides that limit hazards to users, consumers of the products and to the wider environment; support for research on alternative pest management systems and extension of these to farmers. In addition to the correct policies the appropriate actions have to be taken by the various bureaucrats involved in the implementation of the policy.

Improved policy can have an effect. In Indonesia a Presidential order banned 57 insecticides for use on rice following a severe pesticide induced outbreak of brown planthopper. This ban was accompanied by a decrease in pesticide subsidy and increased support for farmer training in IPM. There has been a subsequent 50% reduction in pesticide production and importation while rice yields have increased slightly (Kenmore 1991). This policy was implemented by an inter-sectoral group of ministers that ensured high level support for the IPM program.

Advocacy

In many developing countries, pesticides which have been banned or restricted in developed countries are widely used. In addition, as mentioned earlier in this paper, some pesticides which are approved for use in developed countries are not safe for use in developing countries due to the higher operator exposure rates. Action on these issues can often only be addressed by international groups through pressure on international regulatory authorities and pesticide producers.

A consortium of NGOs including the Pesticide Action Network, Malaysia and the Pesticide Trust, UK are assisting in monitoring and investigating compliance to the FAO Code of Conduct on Pesticides especially the principle of Prior Informed Consent (Pesticide Trust 1989).

The International Development Research Centre (IDRC) of Canada has initiated a dialogue with one of the major manufacturers of pesticides to encourage greater industry responsibility in the developing countries. A "pesticide hazard auditor" funded by industry has been suggested (Loevinsohn 1989). Unfortunately it failed to win the acceptance of the key pesticide industry association. However, IDRC has followed this up with a dialogue involving the key stake holders.

Agroecosystem Controls

There are some cases where major changes have to occur in various aspects of the farming systems over a wide area in order to reduce pest pressure. These may involve management of straw or trash to prevent the carry over of pests or a change from asynchronous planting of rice and vegetables to allow a break in the life cycle of certain major pests. However, these changes require coordinated action by a large number of independent small producers. In some cases these changes have occurred as a result of strong government intervention e.g., synchronous rice planting in Indonesia. In the case of vegetable production in southern China the change from a bureaucratic system (the communes) to a more independent system (the responsibility system) has resulted in a considerable increase in pest management problems due to the break down of an imposed but ecologically sound system of seasonal planting.

Another practice which appears to have the possibility for reduction in pest problems is the integration of rice and fish culture. This traditional practice of growing rice and fish together is expanding rapidly in Asia. Researchers in Thailand, China and Indonesia are all reporting that the presence of fish decreases pests and diseases in rice (Litsinger, Chapter 3). Another interesting factor is that the addition of fish to rice culture appears to encourage farmers to modify their pest control practices and adopt a more judicious approach to the use of chemical pesticides. Thus the addition of fish to rice culture may be one of the first steps in adoption of IPM. However, success of this system in many irrigated areas is dependent on a reduction of pesticides in the irrigation water thus requiring most farmers in an area to reduce pesticide application or use pesticides of lower toxicity to fish.

Alternative Pest Management Tools

Breeding   Plant breeding for insect and disease resistance has been very successful. The International Agricultural Research Centres (IARC's) have had notable successes in major crops (rice, wheat and maize). They have also succeeded in incorporating simple forms of genetic resistance into widely adapted cultivars. This has often succeeded too well, leading to the planting of large areas to only a few varieties. Indonesia grows only 3 varieties on 90% of their rice area and in the early 1980s only one of IRRI's rice varieties was grown on more than 11 million ha across Asia. In these situations when resistance breaks down, the results can be very severe. This has happened in Asia recently with rice blast and brown planthopper on rice.

Biocontrol

Biological control normally involves the release of a predator, parasite, or disease which is then able to control the pest so it causes little damage to the crop. There are three types of biocontrol which show promise for pest management on small farms in the tropics. These are classical biocontrol, innundative releases, and microbial control.

Classical Biocontrol   Classical biocontrol is often effective when pests originally from the center of origin of the crops are accidentally introduced to the crop in its new location. The control involves searching for natural enemies of the pest in its center of origin and then introducing them to the new area. Recently two cassava pests native to South America, the cassava mealybug and cassava green mite have been accidentally introduced to Africa and caused havoc with this crop which has become a staple food for many Africans. In the case of the cassava green mite, *Mononychellus tanajoa* it was accidentally introduced into Africa from South America in 1971 and significantly reduced cassava yields. The control involved a considerable international effort. Natural enemies were surveyed in South and Central America, tested in field conditions in Trinidad, reared in the United Kingdom under quarantine then evaluated in field conditions in Africa and then mass released throughout W. Africa and now E. Africa. There has been considerable success in West Africa.

Classical biological control requires considerable international effort for collection rearing and third country quarantine. Many of the projects work closely with the International Institute of Biological Control (IIBC) in the United Kingdom. Their role has been to assist in the survey for potential controls, taxonomy, determine efficacy and rearing methods, and third country quarantine (Neuenschwander, Chapter 6). The actual release of the natural enemies may not involve local farmers. In the case of the cassava predators the releases were often made by airplane. This often leads to widespread control of a serious pest and has considerable benefit to subsistence farmers.

Inundative Releases   In some cases large numbers of natural enemies are reared and released as a routine control measure. The best example of this inundative release is from China where egg parasitic wasps Trichogramma spp. are reared and released annually to control a number of pests. *T. dendrolimi* and *T. chilonis* are used to control sugarcane borers in Southern China. The control is better, the health problems are much less and the cost is 10% of the previously used chemical control. *Trichogramma* is also used for control of Asia corn borer in the Beijing watershed which formerly was sprayed with DDT.

Microbial Control   There are number of microbial and viral organisms which have potential for use in pest control. A major emphasis has been placed on the development of the bacteria, *Bacillus thuringiensis* (Bt). It appears to have considerable potential as a pesticide because of its safety to mammals and non-target organisms. At least 2.3 tons of commercial Bt products are presently used in developed countries to control insect pests of agricultural crops, forest trees, ornamentals, etc. A very small percentage of this production is used in developing countries. The most important factors limiting its use in developing countries are related to the adaptation of commercial products to local conditions and quality control. There does appear to be potential for local or national production especially when inexpensive local substrates and local Bt strains are used.

Botanical Pesticides

In the tropics there are a large number of plant species with potential insecticidal properties. Many plants are used traditionally for fish poisons, arrow poisons, human medicines and pest control. There are now organizations in the Philippines and Thailand that are surveying promising plants, testing them for insecticidal properties against major insect pests, determining mode of action (specifically looking for new classes of chemicals with novel modes of action), determining chemical structure and in some cases possible synthesis, and determining the effect on non-target organisms and mammalian toxicity. The safest and most promising insecticides are then tested in on-farm experiment for effectiveness under farm conditions. In the future these could then be extracted locally from cultured plants either on a village or homestead scale while some might be commercialized and produced on a larger scale. The hope is to have effective insecticides which can be produced locally, be cheaper than imported products, yet have minimal health and environmental effects.

Integrated Pest Management

Integrated pest management (IPM) is currently the best approach to pest management. It combines a judicious use of chemicals with combinations of the various alternatives mentioned above with various other control strategies. There are, however, many constraints to its adoption (Bottrell, 1987) and it has been adopted in only a few tropical pest management situations. Indonesia is perhaps the best example where IPM has been implemented on rice by a combination of dramatic policy changes related to the pricing and availability of insecticides plus the large scale training of farmers and extension agents (Kenmore, 1991).

Pest Management Decision-Making

Chemical pesticides are often applied on a regular or calendar basis, or as soon as a pest is seen. There is little need for operator decision making. An IPM approach which requires a more judicious use of synthetic or alternative pesticides is based on analysis and decision-making. The farmer must decide when and what to spray based on knowledge of the pest and beneficial insects situation in the crop.

The most common method of determining when to spray is the Economic Threshold Level (ETL). A threshold density of pests are predetermined by researchers. At a pest density below the ETL no economic damage is sustained by the crop where above the ETL crop losses occur. The farmer then has to determine the pest density, normally by counting a set area of the field or by examining a set number of plants and referring to the ETL in order to determine whether to apply pesticide.

The Philippines: A Case Study

The Philippines offers an excellent example of the problems of changing from a chemically based pest management strategy to a more integrated approach which would use fewer chemicals and more alternative approaches.

The Philippines has had a government supportive of this approach since 1986, and there are a number of national and international institutions supporting research on IPM and alternative pest management. This research is carried out by some of the best researchers in the region. The Philippines was the first site of an Inter-country Program for IPM on Rice, and there are a large number of NGO's involved in promoting sustainable agriculture including a number who have been involved in the international advocacy for responsible pesticide use. In spite of all these activities, adoption of IPM by farmers has been very slow. In fact the Philippines still has an agriculture which is heavily reliant on synthetic pesticides. The lack of adoption of IPM may be attributed to several factors including: (1) lack of policy support; (2) lack of local research; and (3) a very weak IPM extension delivery system (Rola and Pingali 1992).

Much of the following consideration of the Philippine situation is based on observations of farmers at two sites, Calamba and Majayjay in Laguna Province, Luzon where rice and vegetables are produced. Additional observations come from two of the authors (Adalla and Rola) who have had considerable involvement in other locations in the Philippines, supplemented by reports from recent publications.

The Current Situation

Health   The health effects of pesticide use continue to be very severe in the Philippines. Marquez et al. (1992) compared health impairments between an exposed group (rice farmers in Nueva Ecija and Laguna using pesticides) and a control group (farmers in Lucban, Quezon who do not use pesticides). Study results showed a significant difference between the exposed and control groups in terms of eye disorders, skin allergies, and seven other internal organ impairments. The health cost function estimated for Nueva Ecija rice farmers showed that as one increases insecticide dose, health costs increases significantly (Rola and Pingali, 1992a).

The frequency of illness among farmers is related both to the amount (shown in Table 2.1) and type of pesticides used (Rola 1989, Rola et al. 1992). Vegetable farmers have the highest frequency of illness, they use the most pesticides, many of which are extremely hazardous (category 1 and 2) including the banned paraquat and DDT. Potato farmers appear to be an exception as their illness frequency is high but their pesticide use is low. However, most potato farmers are also growing other vegetables. Fruit and grain farmers have a similar use rate and frequency of illness with the exception of pineapple growers who use larger quantities of pesticides (mainly herbicides) but of lower hazard and have the lowest illness frequency. Grain and fruit farmers also use the more hazardous (category 1 and 2) pesticides with methyl parathion and monocrotophos being the popular insecticides used by rice and corn farmers.

Farmers probably receive even more exposure to chemicals due to their inadequate knowledge of safe re-entry into the sprayed fields and their ignorance of pesticides labels (Rola et al. 1992). For instance, a measure of farmers' knowledge regarding information on pesticide labels was explored to determine whether the label is an effective strategy for regulation. It was found that most farmers stated re-entry intervals that are much shorter than prescribed in labels. Another survey found that even for farmers

TABLE 2.1 Mean pesticide use (in kg. ai/ha) and frequency of illnesses as experienced by farmers in the Philippines (N = 460).

Crop Cultivated	Mean Pesticide (kg. ai/ha)	Frequency of Farmers Reporting Illness (in percentages)
VEGETABLES		
Cabbage	7.04	56
Onion	3.73	45
Potato	0.96	45
FRUIT		
Mango	2.10	29
Banana	1.77	33
Pineapple	3.84[1]	6
GRAIN		
Rice	1.57	28
Corn	1.16	30
OTHER		
Tobacco	0.43	18

[1]Pineapple farmers use 3.29 kg. ai/ha of herbicides of category 3 and 4.
Source: Rola, et al., 1992.

trained on color coding to rank hazard category, only about 20% answered correctly, while 97% of untrained farmers had incorrect answers.

Philippine farmers use little protective clothing when spraying. Skirts and long pants are the most popular protective clothing while only a few use gloves to protect the hands which receive the highest exposure during the mixing and spraying operations. The backpack sprayers were used for most crops but proper application technologies for tall crops are not available. Hence, the sprayer used for mangoes would consist of a big drum to mix the chemicals and a pump and hose to spray them (Rola et al. 1992). Two persons are usually needed; one pumps the chemical from the drum to the hose; and another handles the hose to spray the mango tree above his head.

Both pesticide storage and disposal practices of small farmers seem to suggest that toxic substances present risks especially to small children. The reason for unsafe storage and disposal practices by farmers may be that they are not aware of the extent of harm that these chemicals can inflict when mishandled; they also cannot afford safe storage and disposal of these materials.

It is obvious that there are severe health effects related to the use of synthetic pesticides by Filipino farmers and the situation has not improved since Loevinsohn (1987) showed increased mortalities among rice farmers due to increased pesticide usage. In fact the implicated pesticide methyl parathion is still widely available. The critical question is why has the situation not improved in spite of an official supportive policy and considerable information on the seriousness of the situation.

Policy and Advocacy

IPM activities in the Philippines started in rice, initiated by the Food and Agriculture Organization (FAO) in the late 1970s. Back-up research was supplied by the International Rice Research Institute (IRRI) and in recent years, the Philippine Rice Research Institute (Philrice) has become active in the research, extension and farmers' training. In May 1987, the Philippine government issued a directive to make IPM the core of the pest control policy in agriculture. This was to be implemented by the Department of Agriculture through efforts of the FAO intercountry program. This was a reversal of previous programs where pest control has been primarily pesticide-based.

There have been conflicting claims on the success of the program. The Department of Agriculture claims a substantial saving of foreign exchange due to decreased importation of pesticides, while the pesticide industry claims there has been no significant reduction in the actual volume of pesticides sold. The few studies aimed at quantifying and measuring the success of the implementation of IPM (Sumangil et al. 1990, Adalla et al. 1990) concluded that Filipino rice farmers are still highly dependent on pesticides. Additional evidence indicates that brown plant hoppers on rice has increased dramatically due to excessive pesticide use (Kenmore 1991).

There is no specific pesticide pricing policy in the Philippines. Pesticide retail prices are significantly affected by taxation, tariffs (import levies) and exchange rates. At the present time, retail prices are lower for older and more hazardous insecticides and farmers use more of these. Newer or less hazardous pesticides such as pyrethroids have higher prices because they are still under proprietary rights. An effective pricing policy would require a selective tax/tariff scheme where less hazardous pesticides are taxed lower than the more hazardous ones.

A regulatory policy is administered by the Philippine Fertilizer and Pesticide Authority (FPA) but implementation has been short of expectation. The local pesticide dealers have demonstrated low compliance with the FPA policy. This is evident by the use of banned pesticides by farmers. In addition the FPA has continue to allow importation of toxic pesticides like methyl parathion. This pesticide is classified as "extremely hazardous". It is banned or severely restricted in a number of countries including

Indonesia, Bangladesh, and Malaysia. Research in the Philippines suggests it cannot be safely used given the current methods of application (Loevinsohn 1989). Its use can be replaced by other less toxic but more expensive pesticides. Yet in 1991 it was the most widely used pesticide among rice farmers in Cotabato Province and widely used on other crops in the Philippines.

The Philippine NGOs are numerous, vocal and often effective at advocacy. In agriculture they have been able to influence official policy on issues of plant genetic resources and biosafety. They have also been active in the international debate on synthetic pesticides but they do not to seem to have been effective in influencing national policy on pesticide use. Loevinsohn (1992) suggests that this is due to the minimal collaboration between researchers and NGOs resulting in the NGO community not having access to the most recent findings, and inadequate technical expertise among the NGOs.

It appears that the national IPM policy has existed only on paper. There has been no high level intersectorial steering committee or coordinating mechanism, no pesticide pricing policy, weak regulation of pesticides and only limited funding for farmer training in IPM (Kenmore 1991). However, because of the problems the Secretary of Agriculture in March 1991 directed his staff to develop the first national pesticide policy statement.

Farm Level Constraints

At the farm level (in Calamba and Laguna), farmers continue to use a number of pesticides which are dangerous to their health. They recognize the dangers, half had experienced some pesticide related effects while family members and neighbors had died of pesticide related illnesses. However, farmers continue to use these pesticides and spray more frequently than required. They are very reluctant to adopt IPM. The constraints to adoption of IPM at this site were:

1. Trust in the value of chemicals and mistrust of the unknown IPM approach,
2. Lack of appropriate tools,
3. Lack of appropriate pest monitoring systems, and
4. A complex household decision making with regard to pest management.

It is useful to examine these constraints in more detail in order to determine more effective ways of implementing an approach that is less reliant on synthetic pesticides.

Farmers Attitude Toward Pesticides In 1973 the Philippine government introduced a rice intensification program, Masagana 99. This program encouraged the adoption of a package of high yield technologies which included the new high yield rice varieties, fertilizer and chemical pesticides. This was supported by a credit program and promoted by an active extension and communication program, and salesmen from chemical companies. The program was successful and resulted in a large increase in rice production. The practice of chemical pest control has been deeply in-grained in the consciousness of the farmers, it has become tradition. This is apparent by the farmers practice of spraying as soon as they see an insect, even though it may be beneficial (shown in Table 2.2). Thus one of the biggest challenges in agricultural development is to help the farmers unlearn what they have learned about pesticide management and practiced since 1973.

TABLE 2.2 Summary of results on farmer respondents' identification of insect pests and natural enemies of rice as an indicator to apply spray. (Total respondents = 45). Source: Rola, et.al., 1988.

Insect	Sprayed (%)	Did not spray (%)	Does not know (%)
Insect Pests			
Brown planthopper	80.0	15.6	4.4
Green leafhopper	84.4	13.3	2.2
Ricebug	75.5	20.0	4.4
Caseworm	86.7	11.1	2.2
Leaffolder	95.6	2.2	-
Whorl maggot	77.8	22.2	-
Armyworm/cutworm	91.1	4.4	4.4
Green-horned caterpillar	80.0	17.8	2.2
Stemborer	84.4	13.3	2.2
Long-horned grasshopper	51.1	46.7	-
Short-horned grasshopper	46.7	51.1	-

Natural Enemies or Beneficial Insects			
Microvelia	55.6	40.0	4.4
Cyrtorhinus sp.	57.8	33.3	6.7
Lycosa	20.0	80.0	-
Apanteles sp.	46.7	46.7	2.2
Beauveria bassiana	82.2	15.6	-
Damselfly	11.1	86.7	-
Coccinellid	64.4	33.3	-

At the two project sites in Laguna the researchers have been successful in reducing the application of pesticides on rice. This has resulted because of a combination of a communication campaign and on-farm participatory research which have demonstrated the benefits of not spraying. A similar situation has occurred at a few IPM project sites elsewhere in the Philippines (shown in Table 2.3). The challenge, however is to achieve much wider adoption of this approach.

The results in vegetables are much different. Farmer cooperators who had seen the benefits of reduced insecticide use on their own fields quickly reverted back to overuse of chemical pesticides when confronted with higher pest levels. In a cash crop like vegetables in which there is a stiff market penalty for damaged vegetables and where farmers often receive credit advances from merchants, farmers are unwilling to risk any crop loss. It appears that a more reliable set of appropriate pest management tools is required for vegetables before farmers are willing to accept IPM.

Appropriate Tools One of the early approaches in the Philippines was to reduce the number of applications of chemical pesticides because there appeared to be no other alternatives to replace an insecticide spray. Moreover, little effort was put into the appreciation of IPM as a site specific management strategy rather than a technology package with broad appli-

TABLE 2.3 Comparison of IPM adopters and non-adopters (Central Luzon and Iloilo, 1989).

ITEM	ADOPTER	NON-ADOPTER	DIFFERENCE
Central Luzon			
Farm size, Ha	2.49	2.58	-0.10 ns
Nitrogen rate, Kg/Ha	70.58	75.24	-4.66 ns
Freq. of insecticide application	1.91	2.36	-0.45 *
Mean yield, Kg/Ha	4146	4015	131 ns
Iloilo			
Farm size, Ha	1.79	2.12	-0.34 ns
Nitrogen rate, Kg/Ha	101	85	16 *
Freq. of insecticide application	2.11	4.44	-2.37 **
Mean yield, Kg/Ha	3647	3547	99 ns

**, * = significant at 5% and 10% levels, respectively.
ns = not significant.

cation that was expected to work throughout the country even though based on only a few experimental sites.

The most recent IPM recommendation for rice calls for the use of recommended resistant high yielding varieties, good water and fertilizer management and limited pesticide use. The decision of when to apply pesticides is based on the preset ETL's for various pests.

Even in rice with considerable research efforts being directed toward alternative pest management there are still no recommendations for alternative pest management approaches. In fact the most serious pest in the Philippines is now the imported golden apple snail (Pomacea). Early control recommendations include the use of TriPhenyl Tin products which have recently been banned from importation and sale. The current recommendations are less pesticide based, but more labor-intensive, techniques such as handpicking, constructing smaller canals and using mechanical barriers in water inlets.

At the Calamba project, the few farmers who applied mechanical control testified that they had eliminated the snail except when the irrigation canals overflowed and migration of old and gravid female snails occurred. In other areas of the country, particularly in Luzon (Northern and Southern parts), where duck raising is a very popular income generating activity, the snail problem is minimized. The ducks are allowed to pasture on rice fields before land preparation and just before transplanting and contribute to the reduction of snails. However, there is a problem of trematodes spread by duck feces which cause skin irritation and itchiness to the rice transplanters. This appears to be an area where an ecosystem approach is required for control which will require collective and area-wide sustained efforts with considerable national level support.

Seeds of the recommended varieties of major crops (rice and maize) are normally available but while most farmers express their desire to obtain the recommended HYV seed, the high cost of certified seeds often prevents them from buying them. Farmers will often obtain seeds from their own field by panicle selecting, from other farmers or from thresher operators who normally sell their share from threshing. Such seed is sometimes claimed to be a new or different variety than what is commonly planted in the vicinity. When farmers save their own seed it is often the women who will assist in seed selection and seed purification.

Alternative technology for pest management in vegetables, a crop that receives heavy pesticide applications is generally not available. Commercially available Bt (Dipel®) has been used for control of Diamond Back Moth (DBM), but farmers in the northern Philippines now report that it is no longer effective. A biocontrol program is currently underway for DBM but the major constraint seems to be the large pesticide applications that kill the released natural enemies.

The research project in Laguna has been exploring various potential controls for the major vegetable pests but the task is formidable. There are at least five major vegetable crops each with its own pest complex, there are few recommended ETL's for vegetables and fewer proven alternative control methods, and there are only a few researchers addressing these questions with little coordination between them. The project is investigating insect parasites and botanical pesticides which are currently available to the project farmers but are not yet widely available to other Filipino farmers.

In conclusion, there are still very few proven alternatives to synthetic chemicals for pest management in the major crops. This in spite of considerable research by international and national research institutions.

Decision Making   The major tool in determining whether a farmer should spray or not is the ETL. Table 2.4 lists the ETLs and methods to determine them for the major rice pests. A recent economic analysis has shown that farmers' practice and natural control are superior to the current ETL's (Rola and Pingali 1992a). This implies that ETL's as recommended by researchers are too low. Further research is needed to develop ETL's that are adaptable to local multi-pest system and other site specific factors. Similar experiences were found at the Laguna sites for rice pest management where it was found necessary to modify the ETLs for site specific conditions.

Two other problems were indicated by farmers. Most of them were in the 40-50 year old range and had failing eyesight. They could not see the pests and natural enemies well enough to make an accurate count. In response to this the project trained school children and unemployed youths to act as pest scouts and do the counting. This worked well as long as the project paid for the scouts, however, farmers were unwilling to pay for them. This may be a result of the perceived cost or it may also reflect the farmers lack of confidence in using hired labor for such an important decision as when to use pest control.

A second problem was the farmers suggestion that the ETL method was inappropriate for their condition. Farmers claim they have their own way of accessing the pest situation. They stand on the bunds and scan the field and determine in their own minds the degree of infestation or damage. The project compared the farmers assessment to that of actually counting pests or damage, in most cases the farmers over-estimated the pest situation. It would seem, however, useful for researchers to further explore methods of insect surveillance that are more acceptable to farmers.

There has been little previous exploration of pesticide decision making at the household level. In the Philippines women generally control the household finances and heavily influence the expenditures on agricultural activities. In Calamba, Laguna farm women would often choose the pesticide but in rice they relied on the advice of the men for decisions on pest management. However, in vegetables where the women are much more

TABLE 2.4 Recommended threshold levels for various insect pest of rice. (Adapted from the DA Interagency Recommendations).

Whorl maggot	2 eggs/hill. Examine 20 hills/paddy twice a week from 7 to 15 days after transplanting. Spray Monocrotophos, like Azodrin, Nuvacron and others, at 0.4 kg ai/ha when ETL is reached.
Defoliators	2 larvae/hill. Examine 20 hills/paddy. When ETL is reached, spray monocrotophos like Azodrin, Nuvacron and others, at 0.4 kg ai/ha.
Stemborers	Examine 100 hills/paddy from 3 to 5 weeks after transplanting up to panicle initiation. Collect the egg masses, place them in a jar and observe daily till the eggs hatch. If more parasites than stemborer hatched from the egg masses, there is no need to spray. If otherwise, spray the field with the appropriate insecticides.
Leaffolder	3 larvae/hill at vegetative stage. 2 larvae/hill at reproductive stage. Examine 20 hills/paddy 15 days after transplanting up to flowering. Spray Azodrin at 0.4 kg ai/ha when the ETL is reached.
Leafhoppers	Use the sequential sampling procedure. List down hopper counts in the corresponding column in the table. Do the same with the friendly insects. Sum up the corresponding running total. Follow the information under the columns of the decision limits.

actively involved in production they normally make the pest management decision themselves. This knowledge is essential in order to effectively target extension advice and training.

It is essential to explore the area of pest management decision making in further detail in order to develop better more adoptable systems and tools in order for farmers to decrease their reliance on synthetic insecticides.

Appropriate Research and Extension   Some of the above problems of lack of appropriate technologies are symptoms of larger problems due to the lack of understanding by researchers of farmers problems. Appropriate pest management will only be developed when there is a close partnership between farmers and researchers using a participatory, on-farm research approach. The research at the two Laguna sites is a start in that direction and offers lessons to future researchers.

Proper dissemination of research results to the farmers is also a crucial responsibility and must be undertaken by well-trained extension agents. Agricultural production technicians (APTs) assigned at the municipal level are the key actors in IPM information dissemination and adoption by farmers in the Philippines. However, they do not have the needed expertise and sufficient resources to bring IPM to the local level (Rola et.al. 1992). They receive little formal training in IPM. APTs claim that constraints to IPM dissemination are lack of logistic support; farmers' negative attitude toward change; improper training of transfer agents; and IPM as a difficult concept. Additional problems are difficulties of visiting farmers related to insurgency problems, accessibility/transportation problems, large numbers of farmers per APT, reorganization and staff reallocation to other groups and too many other activities.

Future Options

Kenmore (1991) argues strongly that the implementation of an effective IPM program that will lead to a more sustainable rice production system will only occur if a number of changes in policy are undertaken together. Using the recent experiences of Indonesia he suggests the following:

- strong government regulation of pesticides based on health and environmental concerns
- removal of government subsidies on pesticides
- increase in the resources available for research, extension and training in pest management
- high level government commitment to achieve success (in Indonesia this consists of an intersectorial ministerial coordinating committee and strong presidential support).
- strong donor support.

In the Philippines some of these policies are in place but implementation has been well short of expectation. This is evident by increased poisoning cases of farmers, a high pesticide residue found in foods taken from farmers' fields, continued pesticide use and a contamination of the natural environment, including groundwater/well-water pollution and pesticide induced resistance.

Rola and Pingali (1992b) have recently suggested a comprehensive strategy to mitigate risks from pesticide use. The strategy suggests the need for:

1. Regulatory Policies,
2. Pricing Policies,
3. Increased investments in training, information and development of application technologies, and
4. Research and extension of alternative pest control measures.

In order to implement these policies there will be a need for strong support from all segments of the agriculture sector, from the government, nongovernment, private sectors and most importantly from the men and women of the farming communities who bear the pain and suffering of the current over reliance on synthetic pesticides.

Acknowledgments

Much of what has been reported in this chapter has been supported both financially and morally by the International Development Research Centre (IDRC) of Canada. Two of the authors (Adalla and Rola) are recipients of IDRC project grants while the other author (MacKay) was a IDRC Senior Program Officer closely involved in the projects. Some of the preliminary thoughts for this chapter were presented at an IDRC sponsored Pesticide Symposium, in Ottawa, September 1990. Thanks go to IDRC for their strong support of these activities.

A special acknowledgement to Dr. Michael Loevinsohn who helped alert the authors and IDRC to many of the serious issues and solutions to overdependance on synthetic pesticides. A recent review of IDRC activities in IPM (Loevinsohn, 1992) also served as a useful guide in developing this chapter.

The contributions of the other researchers and assistants who have been involved in the Laguna research project is also gratefully acknowledged.

Finally the efforts of the small scale farmers in the Philippines who have contributed information, ideas and new approaches is acknowledged. It is they who stand to gain the most from these new approaches.

Bibliography

Adalla, C. B., B. R. Sumayao, M. M. Huque, A. C. Rola, T. H. Stuart, and C. R. Cervancia. 1990. *Integrated Pest Management Extension and Women Project*

(Philippines) Terminal Report (1988–1990). Depart. Entomology. UPLB, 217 p.

Bottrell, D.G. 1987. Applications and Problems of Integrated Pest Management in the Tropics. *Journal of Plant Protection in the Tropics* 4: 1–8

Charbonneau, R. 1989. "Pyrethroid Pesticide: Cotton's Friend, Sprayer's Foe," in *The IDRC Reports*, Volume 18, Number 3, July 1989.

Cheng, E. Y. 1986. "The Resistance, cross resistance, and chemical control of diamondback moth in Taiwan, in diamondback moth management." Proc. First International Workshop, March 11–15 1985. Tainan, Taiwan. pp. 329–345.

Hashim, B. L., and H. F. Yeoh. 1988. "Pesticide Residue Studies in Malaysia," in *Pesticide Management and Integrated Pest Management in Southeast Asia*. Proc. Southeast Asia Pesticide Management and Integrated Pest Management Workshop, February 23–27, 1987. Pattaya, Thailand. pp. 349–354.

Heinrichs, E. A., G. B. Aquino, S. Chelliah, S. L. Valencia, and W. H. Reissig. 1982. Resurgence of *Nilaparvata lugens* (Stål) populations as influenced by method and timing of insecticide applications in lowland rice. *Environmental Entomology* 11: 78–84.

Heinrichs, E. A., and O. Mochida. 1984. From secondary to major pest status: The case of insecticide-induced rice brown planthopper, *Nilaparvata lugens*, resurgence. *Protection Ecology* 7: 201–218.

Ho, T. H. 1965. *The Life History and Control of the Diamondback Moth in Malaya*. Mins. Agric. & Co-operatives. Bulletin 118. 26 p.

Jeyaratnam, J., K. C. Lun, and W. O. Phoon. 1987. Survey of Acute Pesticide Poisoning Among Agricultural Workers in Four Asian Countries. *Bull. World Health Organization* 65: 521–527.

Kenmore, P. E. 1991. *Indonesia's Integrated Pest Management—A Model for Asia*. FAO, Manila, Philippines. 56 p.

Kenmore, P. E., F. O Cariõ, C. A. Perez, V. A Dyck, and A. P. Gutierrez. 1984. Population Regulation of the Rice Brown Planthopper (*Nilaparvata lugens* Stål) within Rice Fields in the Philippines. *J. Plant Protection in the Tropics* 1: 19–37.

Litsinger, J. A. 1991. "Integrated pest management in rice: impact on pesticide." Workshop on Environmental and Health Impacts of Pesticide Use in Rice Culture. IRRI. March 28–30, 1990.

Loevinsohn, M. E. 1987. Insecticide use and increased mortality in rural central Luzon, Philippines. *The Lancet.* June 13, 1987.

Loevinsohn, M. E.. 1989. "Pesticides in the Third World: Controlling Hazards Where Regulation is Weak." Presented, annual meeting of the Agriculture Institute of Canada/Canadian Society for Pest Management, July 10–12, 1989, Montreal, Quebec, Canada.

Loevinsohn, M. E. 1992. *Programme in Integrated Pest Management: Conclusions of a Review and Recommendations for Future Directions*. Natural Resources Division, IDRC, Ottawa.

Loevinsohn, M. E., and James A. Litsinger. 1989. "Time and the abundance of rice pests." Presented, annual meeting Agriculture Institute of Canada/Canadian Society for Pest Management, July 10–12, 1989, Montreal, Quebec, Canada.

Marquez, C. G., P. L. Pingali, and F. G. Palis. 1992. *Farmer Health Impact of Long Term Pesticide Exposure–a medical and economic analysis in the Philippines.* IRRI.

Pesticides Trust. 1989. "Prior informed consent in the international code of conduct on the distribution and use of pesticides." Intermediate Report, Pesticides Trust for the Pesticides Action Network, April 1989.

Ramos-Ocampo, V. E., E. D. Magallona, and A. W. Tejada. 1988. "Pesticide residues in the Philippines," in *Pesticide Management and Integrated Pest Management in Southeast Asia.* Proc. Southeast Asia Pesticide Management and Integrated Pest Management Workshop, February 23-27, 1987. Pattaya, Thailand. pp. 355-367.

Reissig, W. H., E. A. Heinrichs, and S. L. Valencia. 1982. Insecticide-induced arrangements of the brown planthopper, *Nilaparvata lugens*, on rice varieties with different levels of resistance. *Environmental Entomology* 11: 165–168.

Repetto, R. 1985. *Paying the Price: Pesticide Subsidies in the Developing Countries.* World Resources Institute, Washington. 18 p.

Rola, A. C. 1989. *Pesticides, Health Risks and Farm Productivity: A Philippine Experience.* UPLB—APRP Monograph No. 89–01.

Rola, A. C. 1992. *IPM at the Local DA: Results of a Survey.* UPLB/IDRC.

Rola, A. C., R. A. Corcolon-Bagsic, J. T. Hernandez, and A. R. Chupungco. 1992. *Pesticide and Pest Management in the Philippines: A Policy Perspective. An Integrative Report.* UPLB/IDRC.

Rola, A. C., and P. L. Pingali. 1992a. *Pesticides and Rice Productivity: An economic assessment for the Philippines.* IRRI/FAO/WRI (Book in press).

Rola, A. C. and P. L. Pingali. 1992b. "Regulating pesticide use in Philippine rice production: some policy considerations." Presented, Health and Environmental Impact of Pesticides. The Rockefeller Conference Center, Bellagio, Italy, March 30–April 1992.

Rushtapakornchai, W. and A. Vattanatangum. 1986. "Present status of insecticidal control of diamondback moth in Thailand," in *Diamondback Moth Management.* Proc. First International Workshop, 11-15 March 1985. Tainan, Taiwan. pp.307-312.

Soekardi, M. 1988. "Pesticide Residue Control and Monitoring in Indonesia," in *Pesticide Management and Integrated Pest Management in Southeast Asia.* Proc. Southeast Asia Pesticide Management and Integrated Pest Management Workshop, February 23–27, 1987. Pattaya, Thailand. pp. 373–378.

Sumangil, P. Jesus, Arturo J. Dancel, and Romulo G. Davide. 1990. *National IPM in the Philippines.* A Country Report presented, Conference on Integrated Pest Management in the Asia-Pacific Region, Sept. 23–27, 1990, Kuala Lumpur, Malaysia. 54 p.

Tayaputch, N. 1988. "Pesticide Residues in Thailand," in *Pesticide Management and Integrated Pest Management in Southeast Asia.* Proc. Southeast Asia Pesticide Management and Integrated Pest Management Workshop, February 23-27, 1987. Pattaya, Thailand. pp. 343-347.

Teng, P. S. and K. L. Heong. 1988. *Pesticide Management and Integrated Pest Management in Southeast Asia.* Proc. Southeast Asia Pesticide Management and Integrated Pest Management Workshop, February 23–27, 1987. Pattaya, Thailand. 473 p.

Wanleelag, N., and P. Tau-Thong. 1986. *Insect Pests and Residual Analysis of Toxic Substances in Groundnut.* Groundnut Improvement Project. Research Reports for 1982–1985. Kasetsart University, Thailand. pp. 151–161.

Wanleelag, N., and P. Tau-Thong. 1987. *Residual Analysis of the Soil.* Groundnut Improvement Project. Progress report for 1986. Kasetsart University, Thailand. pp. 51–60.

Wanleelag, N., P. Tau-Thong, S. Impitak, A. Sothikul and V. Chawengsri. 1988. *Residues of Heptachlor and Heptachlor Epoxide in Groundnut.* Groundnut Improvement Project. Progress Report 1987. Kasetsart Univ., Thailand. pp. 36–43.

3

A Farming Systems Approach to Insect Pest Management for Upland and Lowland Rice Farmers in Tropical Asia

James A. Litsinger

Introduction

The vast majority of rice farmers in tropical Asia are subsistence farmers: they till less than two hectares and consume most of their production (Barker et al. 1985). Asian rice farmers differ greatly in their adoption of modern crop production methods. Those in irrigated systems with low cost, assured water supply and a favorable rice price can afford to adopt more cash intensive management practices and plant mainly rice. Those in less favorable areas—the rainfed lowlands and uplands—with fewer resources practice more farm enterprise diversity and use fewer agrochemicals.

The mix of pest control tactics for Asian rice farmers varies by environment and farmer resource level. Many of the non-insecticide methods were developed by farmers and are subsumed in the farmers' crop husbandry technology. These cultural methods are effective when performed either on a per field or community-wide basis (Litsinger 1992). Modern high yielding varieties, with a spectrum of pest-resistant traits, are available for most irrigated and favorable rainfed environments, less so in unfavorable rainfed locations. The tropics are rich in entomophagous agents, and natural biological control usually operates at high levels. Insecticides, used mostly by farmers in less risky environments, are at conflict with the activity of natural enemies which are more exposed than the target pest.

The varied pest control options must be integrated into local cropping patterns and management practices, which in the tropics can be highly complex, particularly intensive vegetable/cereal crop systems in the uplands. In one such system, 162 cropping patterns were identified in one village alone (Frio and Price 1978). More complex farming systems occur in the more risk-prone environments and act as insurance against losses due to expected perturbations (Harwood 1979).

The basic unit of integrated pest management (IPM) is the agroecosystem. In the tropics, there are more planting time options, and with small landholdings, farmers spread risk by diversification. This is in contrast with a market oriented farmer who plants only one crop over the entire farm which is managed under an agrochemical blanket of fertilizers and pesticides where relatively few crop management decisions are necessary. Cultural pest control methods are popular with subsistence farmers and their effects are often subtle and best expressed in low energy systems.

Agroecosystems mimic, in community structure, the natural ecosystems they replace (van Emden 1977). Therefore it follows that tropical agroecosystems are structurally complex, as tropical ecosystems are orders of magnitude more complex than their temperate counterparts (Janzen 1973). The principal factor governing complexity is favorable temperature ensuring longer growing seasons, as crop growth becomes limited only by the availability of water.

Numerous studies have shown that direct transfer of agricultural practices (modern cultivars, agrochemicals, machinery) from temperate industrialized countries to the tropics has mostly failed because the ecosystems and client farmers have significant differences in resources, capabilities, and goals (Perrin 1977, Harwood 1979, Gliessman et al. 1981, Altieri 1987). Those same studies, however, showed traditional farmers to be astute and they generally optimized crop production within the resource levels and constraints that they face. Researchers often fail to consider factors other than profit when recommending new technology. Farmers often fail to adopt because of the increased risk or lack of technical knowledge.

More management options are apparent in developing IPM strategies for tropical small-scale systems where cash and land are limited but labor is relatively more abundant. Cultural and mechano-physical control methods will dominate because they require less cash resources but relatively more labor. Longer growing seasons mean potentially more diversified cropping patterns. More labor intensive pest control options mean more methods of control and thus more interactions between pests, crop, environment, and management practices.

As a cereal crop, rice has a relatively high tolerance level for pest damage; therefore the degree of pest control does not have to be as high as that for fruit or vegetable crops with high cosmetic standards. Rice, however, is host to a wide multitude of pests, partly because it is so widely planted. Strong (1979) found a positive relationship between richness of pest species and crop area planted worldwide.

A goal of this chapter is to highlight the farming systems research (FSR) approach as an invaluable method to formulate appropriate insect pest management strategies tailored to equally complex Asian farming systems. The insect pest management strategies are integrated into pest (insect, disease, weed, etc.) management strategies and finally into farming systems management strategies. The diversity of interactions within tropical rice agroecosystems will be emphasized, which on one hand seems highly complex, but on the other hand offers opportunities of integration at new orders of magnitude—another rung in the IPM ladder (Newsom 1980). It is argued that without this further level of integration, pest control technologies will be less acceptable to subsistence farmers.

In the FSR method, focus is on the farmer and his community. Technology generated at experiment stations and farmers' fields can only be validated on-farm. A major tenet in the method is that agricultural technology is a product of the society that invented it (Saint and Coward 1977). Discovery and diffusion of the technology occur in agricultural systems and by involving the farmer early in the developmental process, he will have part ownership of that technology and his eventual adoption is more assured. The most successful IPM programs were created by farmers and researchers (Kenmore et al. 1987).

The technology will be location-specific due to the high interaction between technology and the environment (Zandstra et al. 1981). On-farm testing of the technology assures that it will be tailored to local conditions. The farmers' indigenous knowledge is valued and tested by the team members and farmers themselves when appropriate (Fujisaka 1990, 1991a).

Research is carried out by an interdisciplinary team which describes the recommendation domain (geographic target for the research). Each member sees the problem with an eye from their own discipline and views nature with a different perspective. What emerges is a complement of different perspectives and the team discovers key linkages that exist in the larger system. The creation of a team gives the project more visibility and calls the attention of policy makers who may later provide support.

The farm environment is described through a systems perspective. Most IPM programs are single commodity oriented. As farmers often rotate more than one crop in the same field, a holistic perspective would view pest dynamics and control options for sequences of crops. This holistic approach attempts to understand both technical and socio-economic facets of the existing farming system before making interventions. There is a significant distinction between integrated and holistic. Being integrated is not necessarily holistic. In a holistic viewpoint, the farmers' culture, goals, and constraints become known. In the analysis of the system, the socio-economic milieu is viewed as equally important as the biophysical aspects.

A system performs differently than the sum of its component parts (Ruesink 1976). One cannot predict the outcome even if the system were formulated as a computer model inasmuch as the multitude of interactions, with current knowledge, is unpredictable (van Emden 1982). The FSR method, as opposed to systems modeling follows an empirical approach. The latter can only be effective if the key interactions in the farming system are understood. Researchers should try to understand the system components holistically rather than in a reductionist single discipline fashion of examining only one part while holding the other parts constant.

The team builds changes into the existing farming system. Manipulation should be supported by hypotheses rather than introducing new diversity for its own sake (Litsinger and Moody 1976).

Problems are solved by introducing stepwise changes rather than totally replacing the system. The system is analyzed and interdependencies are described. Technology is tested by trial and error. The farmers are encouraged to carry out their own modifications through farmer participatory trials. There are farmers in each community who continually experiment and their cooperation should be sought (Goodell 1984, Bentley and Andrews 1991).

Problem solving is an iterative and dynamic process. Technology is evaluated by a wider set of criteria other than profit (Farrington 1977). Results are fed back to the experiment station to help redesign new technology that is appropriate to farmers. The farming systems team bridges the gap that exists in developing countries between the farmer, the extension services, and the research station. The FSR approach integrates traditional and modern crop production practices. The extension service becomes a part of the farming systems team, in the process softening the boundary between researcher and extension.

The farming systems team has the responsibility to validate the technology in the target agroecosystem and simplify the recommendations before giving them to extension (Goodell 1984). As recommendations will be made for a target geographical domain rather than for the nation as a whole, they will become more simplified.

The evolution of rice IPM in Asia followed that of IPM in North America. The concept was spearheaded by entomologists in both continents beginning with non-integration where single disciplines focused on single pests. As progress was made, pest management was integrated at more levels (Allen and Bath 1980). For Asian rice farmers, integration of tactics formerly occurred first between host plant resistance and chemical control rather than between chemical control and biocontrol as observed in North America. The beneficial role of biocontrol became apparent later after the phenomenon of resurgence was elucidated (Reissig et al. 1982).

Pest control technology was at first not integrated. Historically, the evolution of pest control technology involved developing methods to determine yield losses and pest status, taxonomy of pests and natural enemies, studies on biology and population ecology, and sampling methods and rearing techniques; all focused on single pests.

Although single tactics were developed against specific pest problems, historically a multitude of cultural and mechano-physical controls developed by farmers dominated (Litsinger 1992). For example, seedbeds were flooded and moats constructed to kill armyworms, and stubble was dug up with special tools after harvest to control stem borers. Use of these tedious methods was minimized with the adoption of modern pesticides (Kiritani 1979). In Myanmar today, with its low labor wages, farmers dig out white grubs by hand from upland ricefields.

Farmers from irrigated fields in Central Luzon in the Philippines perform fewer cultural and superstitious practices (Carbonell and Duff 1980) than those from rainfed wetland (Litsinger et al. 1980, Litsinger et al. 1982) and upland (Fujisaka et al. 1989) environments. Varieties susceptible to diseases were eliminated by farmers by selecting seed from the best growing rice.

Integration of technology occurred at several levels. A review of the array of possible interactions will help in assessing the complexity of the rice agroecosystem, particularly with reference to Asia.

Interaction of Pests and Integration of Tactics Within Pest Control Disciplines

The first level of integration began with trying to kill more than one pest with a single control method. With a plethora of modern insecticides, chemicals began to be evaluated against several pests at the same time during efficacy screening (IRRI 1974). With the advent of the Green Revolution rices, plant breeding programs became sophisticated and cultivars were tested for resistance against several insect pests at the same time (IRRI 1974). Scientists started to integrate components of IPM. Rice varieties were soon bred with resistance to both insects and diseases (IRRI 1974). Insecticide decision making based on economic thresholds embraced more than one pest at a time (Way et al. 1991). As a biocontrol tactic, generalist predators are seen to play a greater role than specialist parasitoids in tropical rice.

Rice insect pest management was first targeted against single pest problems and several tactics (usually chemical control and host plant resistance) were employed against each pest or pest control problem: 1) green leafhopper *Nephotettix virescens*(Distant), vectored tungro disease (IRRI 1973), 2) brown planthopper *Nilaparvata lugens* (Stål) (Oka and Manwan 1978), or 3) against overuse of pesticides and development of insecticide resistance (Kiritani 1977). Components were assembled into technical management packages where farmers grew rice by the numbers, e.g., in sixteen steps for Filipino rice farmers, including four steps to control insect pests (Philippine Ministry of Agriculture 1978).

Interaction of Tactics

Only four insect control tactics are considered, giving six combinations of interactions from examples with rice.

Cultural Control × Chemical Control   Several crop husbandry practices have been employed to improve the performance of insecticides. Harrowing under carbofuran granules during field leveling gave the same level of control of yellow stem borer, *Scirpophaga incertulas* (Walker), and whorl maggot *Hydrellia philippina* Ferino, as did broadcasting the granules into the paddy water at twice the dosage (dela Cruz et al. 1981). Paddy water level can be raised to drive brown planthoppers up the rice plant where foliar sprays can have better contact (Litsinger 1992).

Other cultural control practices act to lower the pest population, which when combined with insecticide application, give better control than either the cultural control or the chemical control method used alone. Lange et al. (1953) report that control of rice leaf miner *Hydrellia griseola* (Fallen) achieved through lowering the paddy water level followed by an insecticide spray was superior to either method acting alone. Early planting reduced gall midge, *Orseolia oryzae* (Wood-Mason) populations, enabling a following insecticide application to achieve better results than later planting (Oka 1983).

Cultural Control × Biological Control   Parasitoids and predators are invertebrates as are insect pests and are often as equally affected by cultural control practices. Cultural control methods often alter the microclimate to the disadvantage of both the pest and its natural enemies (Herzog and Funderurk 1985). Burning rice stubble to kill lingering stem borers is a double-edged sword as it also kills the parasitoids of those stem borers (Lim 1970). Fertilizers may act as a negative factor on the survival of parasitic nematodes of soil pests and stem borers (Prakasa Rao et al. 1975), but fertilizers benefit pests as well as their parasitoids and predators who obtain superior nutrition and attain larger size as a result of feeding on herbivorous hosts growing at a higher level of nutrition (Herzog and Funderurk 1985). For example, pupal parasitoids such as *Tetrastichus howardi* (Olliff) produce more progeny in larger host pupae.

Staggered planting benefits both the pest and its natural enemies (Yasumatsu et al. 1980). The black bug, *Scotinophara coarctata* (Fabricius) female guards its egg mass against parasitoids, but when the field is flooded, she is forced to leave the egg mass, exposing it to the egg parasitoids (Shepard et al. 1988).

Creating refugia for natural enemies to pass the off season is difficult in lowland rice areas which tend to be open expanses broken only by narrow roadways and irrigation canals with little opportunity to support much vegetation.

Cultural Control × *Plant Resistance* The degree of resistance expressed by a variety is related to the pest population level. Cultural control practices can temper pest densities resulting in greater suppression from pest-resistant varieties (Israel 1967, Padmaja Rao 1986). Added nutrients such as silica or potassium can at times increase the level of resistance by hardening culms (induced resistance), while applying nitrogen can render a resistant variety more susceptible (Litsinger 1992). Higher levels of added nitrogen resulted in progressively greater susceptibility in resistant varieties for brown planthopper (Heinrichs and Medrano 1985), gall midge (Chelliah and Subramanian 1972), and yellow stem borer and leaffolder, *Cnaphalocrocis medinalis* (Guenée) (Saroja et al. 1987).

Breeding for early maturity is a pest management tactic as it reduces the time for generational buildup on the crop (Adkisson and Dyck 1980). There is a tradeoff from this approach, however, as early maturing varieties have a shorter period in which to compensate for insect damage, while longer maturing varieties allow more time for pest population buildup.

Plant Resistance × *Biological Control* Combining plant resistance with biological control is usually synergistic and highly beneficial as it reduces the likelihood that insecticides which can upset natural enemy populations will be needed. The combination of host plant resistance and natural enemies gave better control than either tactic alone against brown planthopper (Kartohardjono and Heinrichs 1984) and green leafhopper (Myint et al. 1986) and produced a more favorable pest to natural enemy ratio (Palis 1983). Host plant resistance lowers the pest population to a level that natural enemies can contain. Pests colonize ricefields earlier than natural enemies do (van den Berg et al. 1988), giving them a headstart as often the numbers of natural enemies cannot increase at similar rates.

Qualities of a resistant variety, morphological or chemical, may make such varieties more attractive to natural enemies and increase their tenure on the crop (Herzog and Funderurk 1985). Chemical or morphological factors of resistance may weaken the pest, making it more noticeable and vulnerable to natural enemies.

If the level of resistance approaches immunity, however, natural enemy populations, particularly those of specialists, will suffer from a lack of hosts. Allelochemicals produced by resistant plants may accumulate in natural enemies to their detriment. Pubescent rices may also inhibit natural enemies from locating their hosts.

Plant Resistance × *Chemical Control* The adoption of pest-resistant rices has lessened insecticide usage in Asia to the benefit of the natural enemies (Adkisson and Dyck 1980, Oka 1983). There is no need to apply insecticides against the green leafhopper, the main vector of tungro virus disease, on a resistant variety. On a highly susceptible variety, however, insecticides are usually of no avail; they become effective on moderately resistant varieties (Heinrichs et al. 1986). The additive benefit of host plant resistance and chemical control provides higher levels of control of gall midge than either tactic acting alone (Kovitvadhi et al. 1972).

Resurgence in populations of brown planthopper occurs at a much lower level on resistant than on susceptible varieties (Reissig et al. 1982).

By stressing planthoppers, resistant varieties can lower the LD50 level of insecticides (Heinrichs et al. 1984). The first generation of striped stem borers, *Chilo suppressalis* (Walker) emerging from overwintering sites are more susceptible to insecticides on resistant varieties (Ozaki 1959).

Biological Control × Chemical Control   Because the physiology of insect pests and their arthropod natural enemies are so similar, most synthetic insecticides have detrimental effects on both populations. Injudicious use of insecticides on rice has been the cause of pest population resurgence and secondary pest outbreaks of planthoppers and leaffolders (Litsinger 1989). Insecticide usage on rice in Indonesia has recently been heavily curtailed for that reason (Dilts 1990). Insecticides even adversely affect pathogens of rice insect pests (Aguda et al. 1984).

Careful consideration needs to be taken by farmers before insecticides are used on rice, as insecticide misuse may make matters worse. Some insecticides are less harmful to natural enemies (Fabellar and Heinrichs 1986). Microbial-based insecticides, although not priced economically for rice, offer hope in the future as they are essentially harmless to natural enemies.

Interaction Between Pests   Pests often compete with one another. The presence of one pest may increase or decrease the number of another pest. Examples are given for insect pests, diseases, nematodes, and weeds.

- Insects. Stem borer larvae compete for tillers. Only one larva of the yellow stem borer is normally found per tiller, and larvae of its cohorts or other stem borer species rarely occur together (Rothschild 1971). In West Africa, it was noted that when *Maliarpha separatella* Ragonot is abundant, *Chilo* sp. was not, and vice versa (WARDA 1978). It was also reported that yellow stem borer damaged plants are more liable to injury by black bug (Yusope 1920).
- Diseases. Tungro-infested plants are more resistant to blast caused by *Pyricularia grisea* Sacc. (Singh 1979), but orange leaf virus diseased plants are more susceptible to brown spot fungus, *Cochilobolus miyabeanus* (Ito and Kuribayashi) Dreschsler ex Dastur (Singh 1979). Often, diseases totally overcome a plant and mask the expression of other diseases.

- Nematodes. One nematode species usually dominates in any field in either lowland or upland environments. Generally, the first species to enter the root system destroys the roots before other species can become established (J.C. Prot, IRRI, pers. comm.).
- Weeds. Patches of weeds of a single species are commonly observed and these out-compete and overwhelm other species that may appear. Allelopathy is one mechanism weeds use to ward off other species (Moody 1990).

Interaction of Pests and Integration of Tactics Between Pest Control Disciplines

Pests and their control methods do not act in isolation and may affect each other in ways often unpredictable and counter intuitive (Heinrichs et al. 1986). The greatest number of reported interactions in rice is between weeds and insects, some benefiting pests and some benefiting the rice crop (Moody 1990).

Weeds × Insects

Weed-arthropod interactions are the most common and can be biological or physical in nature.

Biological: With few exceptions, rice insect pests obtain nutrition from plants other than rice (Reissig et al. 1986). Weeds also act as refugia for rice pests from season to season. Such weed hosts harbor the dormant stage of rice pests such as the nymphs of green leafhopper, *Nephotettix cincticeps* (Uhler) or the larvae of gall midge. Insect pests are increased in abundance if the weed serves as a food host between rice crops or during a nonpreferred crop growth stage (eg., preflowering stage for *Leptocorisa* seed bugs [Litsinger 1992]).

If the rice pest selects an alternate host plant during the rice crop, then the pest population is diverted, as in the case of the armyworm, *Mythimna separata* (Walker) in upland rice (Litsinger et al. 1987a) or aster leafhopper, *Macrosteles fascifrons* Stål in lowland rice, where the alternate host becomes a trap crop (Way et al. 1984). Rice pests such as the brown planthopper may oviposit on *Echinochloa colona* (L.) Link, a nonhost. The eggs are inserted into the culms but chemicals in the host are lethal to those eggs (Moody 1990).

Weeds may indirectly affect insect pests by serving as refugia for natural enemies between rice crops. The weedy areas may provide nectar, pollen, or alternate hosts for parasitoids or predators (Barrion and Litsinger 1987). They also offer more structural diversity to spiders and other natural enemies (Bottenberg et al. 1990).

Weeds compete with rice, causing rice to become stunted and chlorotic. Chlorotic plants, for example, are less attractive to green leafhoppers (Khan and Saxena 1985).

Weeds as alternate hosts may serve in a more subtle way by acting to enhance the genetic diversity of the pest species through a broadened host range. This wider genetic diversity, obtained by exploiting more host species, would increase the pest's ability to adapt to population stresses such as insect resistant cultivars or pesticides (Eastop 1981). An example of genetic diversity among rice insect pests is that of *Leersia hexandra* Sw. that serves as an alternate host to a subpopulation of the brown planthopper. The *Leersia*-adapted population of brown planthoppers, however, does not successfully interbreed with the main population (Heinrichs and Medrano 1984).

Physical: On the other hand, weeds may act as a partial physical barrier and prevent natural enemies such as egg parasitoids of stem borers from penetrating a ricefield (Shepard and Arida 1986).

The favorable microclimate in a weedy field of increased humidity and shade is sought as temporary shelter by night flying moths such as stemborers (Moody 1990). Weeds growing in association with rice are damaged by some insect pests, weakening or killing the weeds. The value of this natural biological weed control mechanism was recently quantified (IRRI 1987). Weed weight increase in insecticide-treated plots showed the beneficial effect of insect pests in suppressing weeds. A rice crop under attack by insects is also less able to compete with weeds, giving weeds a growth advantage (Way et al. 1983).

On the other hand, an insecticide protected crop results in better crop growth and therefore less weeds. The outcome is a balance of these three interactions, two of which favor weeds.

Weeding by hand or using herbicide may result in an increase in pest abundance as pest transfer from weeds to rice was recorded in three cases: 1) whorl maggot and caseworm *Nymphula depunctalis* (Guenée) (Moody 1990), 2) armyworm and termites (Litsinger et al. 1987a), and 3) aster leafhopper (Way et al. 1984). Time of weeding determines when the insects will transfer and the extent of damage. Early weeding may kill juvenile insects, whereas later weeding, before the insects pupate, will allow greatest damage to rice.

The herbicide 2,4-D has been shown to increase damage from stem borer (Ishii and Hirano 1963) and gall midge (Sain 1988). The herbicide kills less productive tertiary tillers, forcing damage on the more important primary and secondary tillers.

Drainage is advocated for aquatic rice pests attacking at the vegetative stage, e.g., whorl maggot and caseworm, but this practice encourages weed growth (Moody 1990). The farmers' need to drain their fields to control aquatic insects or the golden apple snail, *Pomacea cannaliculata* (Lamarck) control would be in conflict with their need to pond the ricefield to control weeds.

Thick weed growth inhibits penetration of insecticide and thus prevents contact of spray droplets with the target pest.

Insect pests with urticatious hairs would prevent weeders from pulling weeds. Farmers whose crops are heavily damaged by insects would not find it economical to undertake weeding during such times as a brown planthopper epidemic.

The combination of propanil and certain insecticides has been shown to cause phytotoxicity in rice: propanil and carbofuran (Smith and Tugwell 1975), propanil and m-parathion (Willis and Street 1988), and propanil and CPMC insecticide (Akabane et al. 1968). These insecticides act by inhibiting the enzyme in rice that detoxifies propanil.

Weeds × Diseases

Herbicides have had varying effects on the expression of plant diseases (Moody 1990). IR8 became more susceptible to stem rot fungus, *Sclerotium oryzae*, Catt. when 2,4-D was applied. In another study, MCPA and 2,4-D reduced the severity of bacterial blight caused by *Xanthomonas campestris* pv. *oryzae* (Ishiyama) Dye. Greater infection of sheath blight was observed when butachlor was applied than when no herbicide was used (K. Moody, IRRI, pers. comm.).

Weedy fields alter the crop microclimate making the crop more susceptible to bacterial and fungal diseases which require condensed moisture on plant surfaces to become established. On the other hand, a diseased rice crop is less able to compete with weeds and weed growth would increase.

Diseases × Insects

Insect injury causes wounds in the rice plant, allowing entry of disease organisms. Sucking insects penetrate the plant tissue with their stylet-like mouthparts and introduce disease organisms, particularly during times when insect pests are highly abundant. The following insect and disease associations are chronicled: 1) mealybug, *Brevennia rehi* (Lindinger) feeding and sheath rot caused by *Sarocladium oryzae* (Sawada) W. Gams & D. Hawksaw (Natarajan and Sundara Babu 1988), 2) brown planthopper and sheath blight caused by *Rhizoctonia solani* Kuhn (Lee et al. 1985), 3) planthopper and blast (Kashiwagi and Nagai 1975) and stem rot (Murty et al. 1980), 4) rice bug and bacterial leaf blight (Mohiuddin et al. 1976), 5) tarsonemid mite, *Tryophagus palmarum* (Oudemans) and sheath rot (Rao and Prakash 1985, Chien 1980), and 6) *Maliarpha separatella* stem borer and blast (Pollet 1978).

Viruliferous green leafhopper, *Nephotettix cincticeps* loses its ability to transmit yellow dwarf when parasitized by pipunculids (Santa 1965). A fungicide used to control blast, isoprothiolane, has insecticidal properties against brown and whitebacked *Sogatella furcifera* (Horvarth) planthoppers (Fukada and Miyake 1978).

Nematodes × Weeds

Weeds harbor nematodes making crop rotation as a control measure ineffective if weeds are not also removed (Castillo et al. 1976a). Attempts to control ring nematode, *Criconemella* spp. resulted in heavy growth of yellow nutsedge, *Cyperus esculentus* L. (Hollis and Keoboonrueng 1984). The ring nematode preferred the weed to rice and when controlled unleashed the sedge whose growth had been held in check. Farmers were only able to get higher yields by controlling both the nematode and the weed.

Weedy fields harbor rats and attract birds (Moody 1990). The recently introduced golden apple snail prefers *Azolla* water fern to rice but also feeds on most weeds (Mochida et al. 1991).

Nematode and tungro damage occurring together results in synergistic yield loss (Sarbini and Leme 1987).

Integration of IPM into the Farming System

Early on in the development of IPM, its eventual integration into the farming systems was foreseen (Ordish 1966).

For subsistence farmers, the top-down approach to agricultural development was inappropriate. Researchers in developing countries turned the approach upside down making agricultural development start and end with the farmer (Harwood 1979, Chambers et al. 1989). The FSR approach that was developed had four essential elements (DeWalt 1985): 1) farmer first and last, 2) systems perspective, 3) interdisciplinary team formation, and 4) problem solving process to generate technology which flows from farmer to researcher/extensionist and back.

IPM itself embraces a complex set of site specific technologies and thus is most appropriately dealt with in a farming system perspective. Four examples based on current FSR efforts in Asia follow. These examples show how the FSR methodology was used to develop IPM strategies for subsistence farmers. There are other examples of research teams developing IPM strategies with a FSR perspective (von Arx et al. 1988, Daamen et al. 1989, Fisher 1989, Nwanze and Mueller 1989, Nissen and Elliott Juhnke 1990, Bentley and Andrews 1991).

Irrigated Multirice Cropping Patterns

Developing IPM strategies for rice is aided by understanding what aspects of the farming system favor pest outbreaks. The modern high yielding rice varieties that launched the Green Revolution were photoperiod insensitive and could be grown year round prompting governments to invest in irrigation systems. During the 1960s and 1970s, famine was on the minds of most donor countries for much of Asia since rice yields were increasing at much lower rates than human populations (Barker et al. 1985).

Most rice grown in monsoon Asia was photoperiod sensitive and flowered with short days after the rains subsided. Only one crop was grown per year in more than 90% of ricelands, and during the dry season those ricefields lay fallow. Even though temperatures were favorable, pest development would not proceed due to lack of water and a host plant. A forced off-season was imposed (similar in its pest suppressive effects as a winter fallow in temperate regions) and insect pest and viral disease problems were, for the most part, minimal (Litsinger 1989).

With irrigation, the dry fallow was broken and modern varieties were grown year round. Rice was cropped two to three times per year in many places. Soon a litany of problems began to appear (Litsinger 1989). Viral diseases vectored by insects rapidly increased in incidence and new ones were discovered with almost predictable regularity—ragged stunt, grassy stunt I and II, and wilted stunt (Hibino 1989). Insect vectors—green leafhopper and brown planthopper—reached epidemic proportions. Stem borers changed in dominance from the more polyphagous striped to the monophagous yellow (Loevinsohn et al. 1988). The three main insect pests of irrigated rice—green leafhopper, brown planthopper, and yellow stem borer—are essentially monophagous and benefitted from continuously grown rice. Viral diseases are only perpetuated on living plants.

Weeds were a problem in traditional rice because ponding was not assured. Weeds grew during periods when the rains failed. But cultural conditions with taller traditional rices, in having deeper water (although usually not continuously) and less fertilizer, did not favor weeds. Deeper water caused the selection of the easier to control broadleaved weeds. Modern cultivars are more responsive but poorly adapted to weedy situations, thus it takes more time to weed modern cultivars. Irrigation provided better weed suppression but selected for aquatic species which became difficult to control, particularly where wet seeded rice replaced the more labor intensive and expensive transplanted rice (Moody 1991). In the early vegetative period of wet seeded rice, there is often poorer weed suppression through ponding as the plants are too small and leveling often is not good as deeper water levels would kill young plants in low-lying areas. Rotating an upland crop with wetland rice resulted in a shift in weed species and reduced weed problems (De Datta and Jereza 1976).

Rat problems also became worse (Lam 1991). Ricefield rats have a high reproductive potential and adjust their breeding cycles to the available food. Rats are adapted to aquatic conditions and year round cropping with irrigation allowed them to breed continuously.

Continuous cropping favors not only viral diseases, but fungal diseases as well (T.W. Mew, IRRI, pers. comm.). Fungal inoculum of endemic sheath blight and stem rot diseases increase proportionally with rice cropping intensity. Fungal pathogens that attack the grains and cause dirty panicle disease also increase. Red stripe, a new disease in Indonesia and Vietnam, is only found in intensively cropped areas.

Rice nematodes are found wherever rice is cultivated (J.C. Prot, IRRI, pers. comm.). The greatest number of nematode species that attacks rice occurs in the uplands. In the Philippines, the dominant upland rice nematode is *Pratylenchus zeae* Graham. The rice root nematodes *Hirschmanniella* spp. are adapted to flooded conditions and are most prevalent in lowland rice. The rainfed wetlands, which form the largest hectarage of rice in Asia, represent an intermediate situation as half of the year is wet and the other half is dry, favoring neither *Pratylenchus zeae* nor *Hirschmanniella* spp. Some nematodes, however, always survive to attack the crop. As the rainfed wetlands became irrigated, fewer nematode species could adapt to the flooded conditions, but those that could survive thrived. *Hirschmanniella* spp. became even more prevalent when rice was double or triple cropped. *Hirschmanniella* spp. populations increased in relation to the increased yield as higher yielding plants, made possible by irrigation, have larger root masses.

Soil physico-chemical properties change under prolonged flooding (De Datta 1981). Even though soil nitrogen and carbon progressively increased over a number of years under anaerobic flooded condition, response to nitrogen fertilizer decreased as the soil increasingly held on to nitrogen, chemically rendering it unavailable to the plant (Cassman et al. 1992). Prolonged flooding also exacerbates zinc deficiency in rice soils.

Analyses of long-term irrigated rice cropping trials showed a progressive yield decline in both fertilized and unfertilized plots, regardless of whether nitrogen was applied from organic or inorganic sources. The result from the experiment station at IRRI (Flinn and De Datta 1984) is not an isolated phenomenon and has been corroborated in numerous other locations in Asia. The decline has been attributed to the problems arising from prolonged ponding.

Under intensive management, short-season rices can be grown in a continuous cycle. Corporate farms established during the 1970s in the Solomon Islands and the Philippines harvested up to four crops per year from the same field (MacQuillan 1974, Loevinsohn 1991). By the early 1980s, all were abandoned because of uncontrollable rat, insect pest, and disease problems resulting from sustained favorable conditions for pest multiplication.

Insect pest and viral disease problems were additionally favored by injudicious use of insecticides which caused planthopper and leafhopper resurgence (Reissig et al. 1982). An answer to the problem has been to recommend a rice-free period during the cropping pattern which removes rice from the landscape for several months each year and dries the soil (Loevinsohn et al. 1988, Litsinger 1989). Two crops of early or medium maturing modern rice can be grown per year and this would allow 2–3 months for a third upland crop such as a grain legume or vegetable. The timing of the upland crop should preferably be during the dry season. It can even be irrigated as long as the fields are not in standing water. The break in the rice crop during the driest parts of the year conserves water and also dries up the soil, causing volunteer rice to wither. This is a preventative measure, which in ecological terms, reduces the carrying capacity of the rice environment to pest populations.

There is a symbiotic relationship between wetland rice and dryland crops, particularly noncereals. The flooding during the rice crop also offers pest suppression to the upland crop. Upland nematodes (*Rotylenchulus reniformis* Linford & Oliveira and *Meloidygne incognita* [Kofoid & White] Chitwood) are killed off (Castillo et al. 1976b) as are soilborne bacterial organisms such as those which cause bacterial wilt *Pseudomonas solanacearum* EFS of tomato. The rice crop therefore breaks the cycle of viral diseases, insect pests, and weeds of upland crops.

Standing rice stubble offers protection against early-season insect pests of cowpea and mungbean if the crop is established by minimum tillage, a popular technique among farmers (Litsinger and Ruhendi 1984). The standing rice stubble interferes with the colonizing ability of thrips (*Thrips palmi* Karny), leafhoppers (*Amrasca biguttula biguttula* [Shiraki]), and aphids (*Aphis craccivora* Koch).

Beanflies (*Ophiomyia phaseoli* [Tryon]) land on the tall stubble when the legume crop is small, causing many of them to fly out of the field. Protection is offered from beanfly attack when the rice stubble is taller than the legume. Beanflies land on the crop canopy seeking the youngest leaves. When they land on the stubble and not the host, many fly off seeking a better habitat.

Management practices of farmers at four irrigated sites in major rice-growing regions of the Philippines were monitored over a number of seasons. Disparities between the researchers' recommended practices and those of the farmers became evident (as shown in Table 3.1). In particular, farmers planted more seedlings per hill than what researchers recommend. Research at IRRI showed 2–4 seedlings per hill to be sufficient and any further increase was wasteful.

Taking a cue, seedling number per hill was tested and related to the crop's ability to tolerate insect damage. It was discovered that increasing the number of seedlings from 3 to 12 seedlings per hill on modern high-tillering rices provided the ability of the crop to compensate for stem borer damage levels up to 2% whiteheads (as shown in Figure 3.1).

Additionally, it was noted that farmers often applied nitrogen fertilizer

TABLE 3.1 Comparison of researchers' and farmers' crop husbandry practices for rice in an irrigated wetland site. Guimba, Nueva Ecija, Philippines, 1984-91.

	Researchers' practice	Farmers' practice
Seeding rate		
Transplanted crop	50 kg/ha	149 ± 10 kg/ha
Direct seeded	100 kg/ha	209 ± 7 kg/ha
Land preparation	1 plowing	1.3 ± 0.1 plowings
	1-2 harrowings	2.8 ± 0.1 harrowing
Age of seedlings (days after sowing)	21 days	3.1 ± 0.8 days
Nitrogen application		
Seedbed	None	36 g/m2
Field Wet season	70 kg/ha	124 ± 15
Dry season	50 kg/ha	84 ± 7
Insect pest assessment	Quantitative	Qualitative
Insecticide application (no.)	Economic thresholds	1.4

at higher rates than seemed optimal. In Guimba, Nueva Ecija, Philippines, for example, farmers applied more than 100 kg N/ha while researchers found the optimal rate to be 70 kg N/ha. This again provided another cue and again an increasing ability of the crop to tolerate stem borer damage at higher nitrogen rates was found (as shown in Figure 3.2). Yield loss from increasing levels of stem borer whiteheads occurred in a linear fashion without added nitrogen fertilizer; but it had a curvilinear relationship with added fertilizer showing compensation up to 3% whiteheads.

More time is needed by the crop to outgrow pest damage and very early maturing varieties such as IR58, a 90—day variety, was noticed to have a higher yield loss than longer maturing varieties (Litsinger et al. 1987b). A trial compared the yield loss of IR58 with that of IR74, a 125—day variety. Yield loss was determined as the difference between yield of insecticide protected plots and those of unprotected plots for each variety. A significant yield loss of 0.8 t/ha occurred on IR58; IR74 had an insignificant loss of 0.2 t/ha (as shown in Table 3.2).

The disparity between researchers' and farmers' practices can be explained by the reductionist manner researchers at experiment stations use to test the various crop management components. Researchers test one component at a time while holding all others constant. In testing the optimal number of seedlings per hill the researcher grew the crop under optimal conditions and applied a blanket of insecticide protection throughout the course of the trial. The conclusion was that 2–4 seedlings per hill gave optimal yield. Increasing the number of seedlings under those conditions was deemed unnecessary. However, the results can only hold true in the absence of insect damage.

Similarly, researchers formerly noted that increasing the rate of nitrogen application leads to increasing numbers of insect pests; lower rates were therefore recommended (Litsinger 1992). Our research concurred with those findings but found that the crop gave a higher yield despite sustaining greater insect damage.

A nitrogen-fixing, flood-tolerant, green manure crop such as *Sesbania*

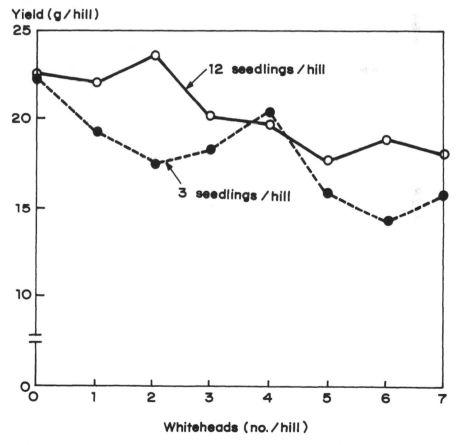

FIGURE 3.1 Greater tolerance for rice stemborer whitehead damage with higher seedling density per hill. Zaragoza, Nueva Ecija, Philippines, 1989 dry season.

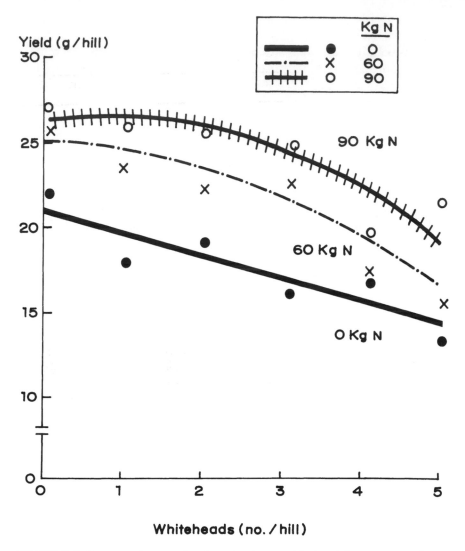

FIGURE 3.2 Greater tolerance for rice stemborer whitehead damage on a nitrogen (N) fertilized crop. Zaragoza, Nueva Ecija, Philippines, 1989 dry season.

rostrata Brem. can be considered before or between two rice crops, particularly if there is a period of heavy rainfall which is not conducive to rice establishment. A beneficial effect is that an organic nitrogen source, releases nitrogen more slowly, allows greater tolerance of vegetative damage occurring singly or in combination (as shown in Figure 3.3). In the experiment, *Sesbania* was plowed into the soil after 45 days of growth. The nitrogen content of the plant biomass was estimated to be 98 kg N/ha. Tolerance to whorl maggot and early season defoliating worms (*Naranga aenescens* and

TABLE 3.2 Comparison of two varieties with different growth maturities on tolerance for yield loss from insect damage.a Zaragoza, Nueva Ecija, 1990 WS.

	Yield (t/ha)	
	IR58 (90 days to maturity)	IR74 (125 days to maturity)
Insecticide protectedb	5.1 a	5.2 a
Untreated	4.3 b	5.0 a
Difference	0.8** 16%	0.2 ns 4%

ns = not significant at 5% level.
**significant at 1% level.
a. Average of four replications. In a column, means followed by a common letter are not significantly different by LSD.
b. 9 insecticide sprayings.

Rivula atimeta) was greater than that in adjacent plots where 80 kg N as inorganic urea was applied. Yield however in the plots with *Sesbania* incorporated were lower than those with inorganic nitrogen, due to the slow release of nitrogen. Yields should be comparable if the practice is repeated annually.

Sesbania offers an additional benefit in nematode management as damage in ricefields is minimized in the following rice crop (Prot et al. 1992).

The ability of a rice crop planted to a modern variety to tolerate insect pest damage is usually less when several pests occur at the same time. Combinations of pests damaging the crop often produce synergistic losses in spite of cultural management (Litsinger 1991). A crop can tolerate damage by whorl maggot as a single pest. In fact, whorl maggot damage at IRRI is so readily tolerated that researchers were doubting its pest stature (Shepard et al. 1990). In Nueva Ecija, Central Luzon, however, the same level of whorl maggot damage, in conjunction with defoliation by caterpillars, caused high yield losses (Figure 3.3). Defoliators are more abundant in Nueva Ecija, a more extensively cropped area, and a level of 19% damaged leaves is normal. In contrast, the level of damaged leaves from vegetative stage defoliators averaged less than 2% in Laguna, Philippines, a much smaller rice area.

A trial in Nueva Ecija compared whorl maggot damage alone and in combination with defoliators and defoliators plus weeds at four levels of nitrogen fertilization. There was no yield loss from whorl maggot damage alone at any nitrogen level (Figure 3.4). In combination with defoliators,

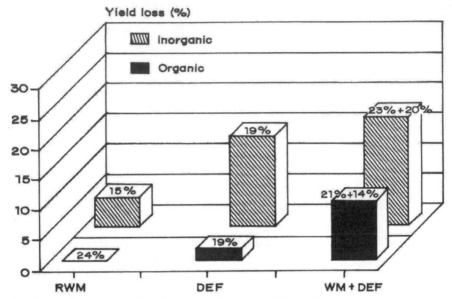

FIGURE 3.3 Rice yield loss from whorl maggot (*Hydrellia philippina*) and vegetative stage defoliators (*Naranga aenescens, Rivula atimeta*) alone and when combined at two fertilizer regimes: inorganic (80 kg urea/ha) and organic (98 kg N/ha as *Sesbania rostrata* green manure). Values are percentages of damaged leaves from each pest. Guimba, Nueva Ecija, Philippines, 1990 wet season.

there was a high yield loss (24–36%) at low nitrogen levels but the loss decreased to less than 5% at 90 kg N/ha. In combination with weeds, however, high yield losses occurred (32–42%) at any nitrogen level because weeds were highly competitive.

From our knowledge of whorl maggot and defoliators, if both were present, the farmer would have the choice to control but one of the pests. As it is more difficult to control whorl maggot, the farmer should target his efforts against defoliators. Other pest combinations need to be to determined to identify problems that farmers can ignore, saving their scarce resources. From the example in Figure 3.4, weeds are more important to control than whorl maggot or defoliators but defoliators plus whorl maggot can be tolerated at high nitrogen levels. Stresses other than pests also influence the ability of the crops' physiological processes to compensate and should also be considered (Litsinger 1991). Highest losses from insect pest damage usually occur on a poorly growing crop suffering from stresses such as drought or zinc deficiency (Litsinger et al. 1987b).

In light of this knowledge, when farmers' fields are monitored for pests, less time should be spent counting densities of each pest but more time

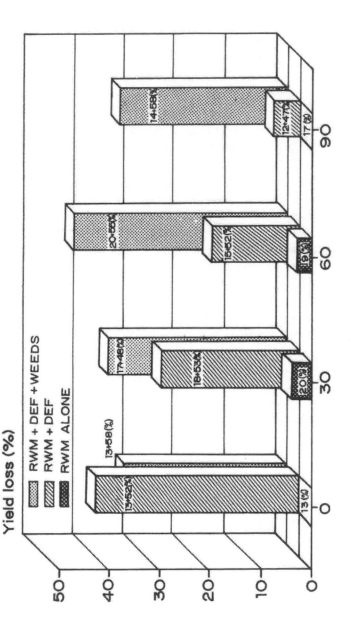

FIGURE 3.4 Yield loss from combinations of whorl maggot, insect defoliators (*Naranga aenescens* and *Rivula atimeta*), nitrogen stress, and weeds. The values on the histogram bars are the percentage of damaged leaves from whorl maggot and defoliators, respectively. Zaragoza, Nueva Ecija, Philippines, 1991 dry season.

should be devoted to recording which pests and stresses are present at levels likely to cause economic loss when occurring together.

When farmers' practices were monitored in Zaragoza, Nueva Ecija for six consecutive crops from 1984 to 1987, the bases of insecticide application decisions were classified into three categories: 1) presence of insect pests in the field, 2) evidence of insect pest damage, and 3) calendar-based scheduling. During the 1984 wet season crop, 19% of decisions were preventative as the rest of the farmers were willing to risk losses from insects when they based their decisions on the presence of pests or damage (Table 3.3). An outbreak of tungro occurred during the 1984 crop and, consequently, a greater number of decisions (33%) which were preventative in nature were made during the following crop. This time farmers were more concerned with having enough rice to eat than with making high profits.

Drought occurred during the 1985 dry season which again affected the farmers' decision-making process in favor of prophylactic applications (this reached 44% in the 1985 wet season crop). During the 1985 wet season, however, a typhoon caused great damage to the standing crop and farmers' decisions to use prophylactic applications reached 46% during the 1986 dry season crop as farmers were again concerned with adequate supply than with profits. Yields were good during the next three crops and more farmers became profit conscious; decisions regarding prophylactic applications were reduced from 46 to 7–26%. This example shows that the subsistence farmers' bases for decision making can change and that new technology must be adapted to these often-changing conditions.

The IPM team will continue to look for low-cost methods of insect pest management both by identifying cultural practices to help the crop tolerate damage and by estimating yield loss from combinations of stresses. The insect pest control strategy will be based on the combination of stresses at any given growth stage.

Rice-Fish Culture

Rice and fish make up the basic diet of the average Asian. In Asia, the price of rice usually favors the urban consumer, not the rice farmer. Consequently, there is limited income earning capacity for farmers tilling only rice. The degree of enterprise diversity is also limited with irrigated rice farmers. An exception is to raise fish. Rice-fish culture is a relatively simple technology with low capital outlay and quick turnaround, but it requires good water supply. This enterprise uses scarce agricultural water and land more optimally, resulting in a symbiotic relationship where both rice and fish benefit.

Rice-fish culture began with farmers trapping fish in their ricefields and eventually evolved into stocking the ricefields soon after transplanting to simultaneously raise fish (Spiller 1986). The observation was that fish grew faster in ricefields than in ponds (Ardiwinata 1957).

TABLE 3.3 Farmers' reasons for applying insecticide to rice in Zaragoza, Nueva Ecija, 1984-87.

Reason	Responses (%)						
	1984 WS	1985 DS	1985 WS	1986 DS	1986 WS	1987 WS
Presence of insect	41	27	1	29	31	53
Presence of insect damage	37	39	53	25	43	39
Prophylactic	19	33	44	46	26	7
Motivation	Profit	Food	Food	Food	Profit	Profit
Production problem	Tungro	Drought	Typhoon	None	None	None
n	60	35	41	34	41	30

Farmers were interviewed during the crop season and their answers as to why they applied insecticide were categorized. The motivation to produce more rice for food (apply prophylactic scheduling) rather than maximize profit in the current crop was determined by production problems during the previous season.

Raising fish in ricefields has been traced to the Eastern Han Dynasty (25–220 AD) in China, thriving until the 1950s when petroleum-based pesticides were widely used (Li 1988). Rice-fish culture, however, declined throughout the rice growing world with the heavy use of organophosphate and organochloride insecticides. Farmers were further discouraged by the early maturity of the new rices as fish did not have time to grow to market size.

In the Central Plains of Thailand, pesticide use led to a dramatic decrease in rice-fish culture (Fedouruk and Leelapatra 1985). For the same reason, rice-fish culture declined in Java, Indonesia (Koesoemadinata 1980) and Malaysia (Moulton 1973). Taiwanese and Japanese rice farmers were also heavy users of pesticides and fish production in ricefields was essentially eliminated in the 1950s and 1960s (Chaudhuri 1985).

Heavy insecticide usage in rice occurred as a result of government programs which subsidized pesticides and offered inexpensive credit to farmers. From the viewpoint of these government agencies, insecticides were treated like fertilizers because they were thought to be needed each crop to attain high yields (Kenmore 1987, Kenmore et al. 1987).

Pesticides affect fish both directly (cause death and suboptimal growth) and indirectly (kill fauna in the food chain and accumulate as residues and thus affect man, the end consumer). In Bangladesh where fish contributes 80% of animal protein in the average daily diet, fish consumption decreased from 33 kg per capita in 1963 to only 21 kg in 1983, partly due to a decline of rice-fish culture (Ahmed et al. 1992).

With the development of pest-resistant varieties and the adoption of IPM, rice-fish culture is being revived in Indonesia, Thailand, and China. In China, nearly a million hectares are in this system, the average annual fish yield being 300 kg/ha (Xu and Guo 1992). Indonesia has more than 100,000 ha in rice-fish culture and averages more than 500 kg fish/ha with no apparent decline in rice yields.

Factors that evolved with modern varieties also favored rice-fish culture. More irrigation systems allowed more farmers to consider rice-fish culture. Photoperiod-insensitive varieties allowed rice to grow year round with a potential of two fish cycles per year. Increased fertilizer usage meant greater production of fish food in ricefields.

The pesticides now recommended for rice are less toxic to fish. Adoption of rice-fish culture would also mean fewer reasons to use insecticides in rice. If fish control some rice pests and raise rice yield, then there is less cause to apply insecticides.

Research reports claimed 7–14% increase in rice yield over a sole crop of rice even though 10% of the plants were removed for trenches (Spiller 1986). Three reasons were proffered as explanations for this phenomenon (Spiller 1986, Lightfoot et al. 1992):

1. Fish feces provide a new source of nutrients for the rice crop. A carp produces 36 kg of fresh feces in its lifetime. Fish also eat algae that normally lower the pH which in turn reduces NH3 losses. The paddy soil N,P,K content increases with rice-fish culture compared to rice alone,
2. Movement of the fish stirs the soil, providing aeration that improves nutrient availability and reduces nitrogen denitrification losses. Tillering is also believed to be enhanced by the activity of the fish, and
3. Fish act as biological control agents to lessen pest incidence.

Even higher increases (10–20% more) have been attributed to improved management practices because farmers are more cognizant of the presence of fish and want to secure their investment. Better water control practices in turn lead to increased yield independent of the presence of fish. Deeper ponding, for example, ensures better weed control and floods rat nests. For many ricefield weeds, germination, emergence, and growth are reduced with increased ponding levels. Rats make their burrows in rice bunds adjacent to the field.

Fish eat weed seeds and emerging weeds in ricefields (Cagauan 1991, Moody 1992). Grass carp are particularly good herbivores and a density of 2 grass carp per m2 results in the elimination of most weeds (Nie Dashu et al. 1992). Rice plants will also be eaten if they are too young. Because the weeds are grazed as they emerge from the soil, few nutrients are lost to the rice. The degree of weed control, however, will vary with weed species, stocking rate, and species of fish. Increased water depth in rice-fish fields will also result in reduced weed growth.

Insect pests and diseases are reported to be controlled by fish. The Chinese have done the most research in this area. Hora and Pillay (1962) cite less yellow stem borer damage in rice-fish culture. Xu and Guo (1992) reported that the presence of carp resulted in ten times fewer planthoppers, three times more spiders, and one third less incidence of rice stripe disease. Rice stripe virus is vectored by the smaller brown planthopper, *Laodelphax striatellus* (Fallen).

In another study, the presence of fish reduced brown planthopper populations by 40%, yellow stem borer deadhearts from 0.3 to 0%, leaffolder damaged leaves from 4.4 to 2.2%, and sheath blight from 80 to 58% infected plants (Xiao 1992). Liao (1980) stated that fish culture resulted in a reduction of stem borer egg masses from 1200 to 300–900/ha, deadhearts from 0.5 to 0.1–0.3%, whiteheads from 0.5 to 0.25–0.3%, leaffolder-damaged leaves from 50 to 12-44/100 hills, brown planthoppers from 8 to 3–5/plant, and green leafhopper from 8 to 2–6/plant. Li (1988) reported that fish reduced *Chilo suppressalis* stem borer deadhearts by 53% and fed on planthoppers (*Nilaparvata lugens, Laodelphax striatellus*), green cater-pillar, *Naranga aenescens*, and snout beetle, *Echinocnemus squameus* Billberg which fell in the water. In India, populations of yellow stem borer were less with fish than without (Datta et al. 1986).

In a study in Northeast Thailand with tilapia, carp, and silver barb reported that three rice diseases occurred less frequently in rice-fish culture: 1) narrow brown spot caused by *Cercospora oryzae* Miyake, 2) sheath blight, and 3) bacterial leaf blight (MacKay et al. 1988). In the same study, leaffolders were fewer, whorl maggot damage was greater in the presence of fish, and gall midge damage levels were minimal.

Fish are recommended as biocontrol agents against the golden apple snail in Japan and Taiwan (Mochida 1991). The fish feed on the juvenile forms.

Several pest control mechanisms involving fish are offered. Nakasuji and Dyck (1984), in a greenhouse study, found that up to 67% of the brown planthopper population dropped from host plants in one day. A greater number of older nymphs and adults fell compared with younger nymphs. These averages will probably be higher in the field, especially on windy days.

Stem borers, which lay eggs in masses, normally disperse aerially (as neonate larvae dangling from silk), then drop to the water surface. Upon hitting the water surface the shock wave produced alerts the fish. Aquatic predatory insects would also fill this role, but they may also be consumed by the fish. Fish may simply be better predators. Chapman et al. (1992), however, reported more spiders in rice-fish culture, which presumably included water surface-dwelling wolf spiders, the most common species in wetland rice culture. An unconfirmed mechanism reported from the Sukamandi Rice Research Station in Java, Indonesia, is that of rice leaffolder control by fish activity—their splashing causes the moths, which reside in the field during the day, to fly (A.M. Fagi, SURIF, pers. comm.). Birds and dragonflies then prey on them. The moths are normally active at night when predator activity is less.

It is presumed that sheath blight is reduced because fish eat the sclerotia resting body, thus reducing the inoculum load in the field. Changes in soil fertility due to the presence of fish may have influenced incidence of narrow brown spot and bacterial leaf blight diseases.

There is no basis for governments to subsidize insecticide usage in rice as studies have shown that losses are more variable field to field than season to season or year to year (Kenmore et al. 1987, Litsinger et al. 1987b). High field to field variability in losses means insect populations are unpredictable; fields should therefore be monitored for pests on a regular basis. Corrective rather than preventative action has been shown to be most profitable (Smith et al. 1989).

Attainable yield losses in four locations in the Philippines as determined by the insecticide check method (Litsinger 1991) range from 0.30 to 0.76

t/ha or 19% (as shown in Table 3.4). Farmers, however, are unable to prevent these losses with their level of insecticide technology, and unfortunately there is little scope for farmers to improve with current technology. Farmers' yields in the four locations, where the number of sprayings ranged from 1 to 2 per crop, were no different from an untreated check.

The main problem with farmer's application is underdosage (Litsinger et al. 1980). Filipino farmers overwhelmingly prefer sprayable insecticide formulations over granules. Few farmers can afford a motorized, forced-air sprayer that gives good coverage. Using the lower cost knapsack sprayer with low pressure gives poor insecticide penetration through the canopy of a dense crop such as rice. Farmers are hesitant to increase the dosage per tank load as they fear for their health (Goodell 1984). Also, they are not willing to increase the number of sprayerloads per hectare and change their nozzle that delivers finer spray droplets and more thorough coverage. Therefore, the only conclusion is that farmers should not apply insecticides unless they can increase the dosage through better coverage.

The added risk of harming fish with insecticides needs to be considered by farmers even though techniques of minimizing insecticide exposure have been developed. The preplant method of soil incorporating a systemic granular formulation is unacceptable in pest management terms as this is a prophylactic treatment and losses cannot be predicted before planting (Estores et al. 1980). Another method is practiced in China—fish are driven into trenches by lowering the water level before spraying (Li 1988). Water is returned to the ricefield after several days. Insecticides less toxic to fish are used (Koesoemadinata 1980). Safer insecticides are being developed but, being new chemicals, will be more expensive.

The rice-fish team will continue to encourage fish raising by small-scale rice farmers to increase their incomes. The 7–14% higher yield claimed as a direct result of the presence of fish in part compensates for the 5–19% attainable yield loss recorded in rice from insect pest damage (Table 3.4). It remains to be seen if losses from insect damage in rice-fish culture could be lowered, in effect removing the need for most insecticides. Rice-fish culture is a good incentive for farmers to use less chemicals and the technique is being introduced as a part of IPM extension programs which discourage farmers from using insecticides.

There is a symbiotic relationship between rice-fish culture and IPM. IPM paves the way for rice-fish culture and rice-fish culture is a reward for farmers that adopt IPM.

Rice-Wheat Cropping Pattern

The yield of both rice and wheat from the rice-wheat cropping pattern planted in 11.6 million hectares in the Himalayas region of Pakistan, Nepal, India, and Bangladesh is stagnating in relation to the need to feed the growing populace (IRRI 1990, 1992; Macklin and Rao 1991). These

72

TABLE 3.4 Yield losses in rice from insect pests measured by the insecticide check method compared with the yield gain from the farmers' insecticide use in the Philippines.

| Site | | Years | No. Crops | Yield loss[a] | Farmers' practice | | |
Town	Province			t/ha	%	Yield gain over untreated (t/ha)[b]	Insecticide applications (no.)
Calauan	Laguna	1984-90	12	0.36*	5	-0.13ns	1.3
Zaragoza	Nueva Ecija	1988-90	3	0.30*	11	0.01ns	1.9
Guimba	Nueva Ecija	1984-90	13	0.77 ± 0.07**	19	0.20ns	1.4
Koronadal	South Cotabato	1986-90	13	0.76***	17	0.28ns	2.1
Mean				0.54	13	0.09	1.8

aYield difference between insecticide treated (9 applications at high dosages of the most effective insecticides) and untreated plots.
*P < 0.05, **P < 0.01, ***P < 0.00.
bns = P > 0.05.

two crops supply 50–80% of the calorie intake of the people in these four countries.

For example, yield of rice in the Tarai belt of Uttar Pradesh, India, increased by 47% (1.9–2.8 t/ha) from 1979 to 1989, while that of wheat correspondingly rose by 53% (from 1.7 to 2.8 t/ha) (Hobbs et al. 1991). To achieve these yields, farmers' rates of nitrogen fertilizer application rose by 97% (from 58 to 114 kg N/ha) over the decade.

Rice and/or wheat yields in the rice-wheat rotation are declining in some areas (Hobbs et al. 1991, Macklin and Rao 1991). In most of the region, wheat is a new crop where formerly only a single, low yielding rice crop was cultivated. In a long-term yield trial spanning 16 years (1973-89), rice yield in the nonfertilized plots decreased from 5.5 to 3.5 t/ha, a 36% reduction (as shown in Figure 3.5). Yields also declined in the plots receiving high levels of inorganic (N P K) and organic (farmyard manure) fertilizer from 8 to 6.5 t/ha, a 19% decrease.

Traditionally, farmers fertilize their fields with manure from milking carabaos and draft bullocks. Increasingly, deforestation is making the supply of fuel wood more expensive and farmers compensate for this by using dried manure cake. Housewives prefer dried manure cakes for cooking as they burn clean compared with the smokey forest wood. Therefore less manure is available as fertilizer. In the favorable parts of the tarai, farmers purchased tractors which replaced the bullocks, making even less manure available. To maintain yield levels, farmers were forced to buy inorganic fertilizer.

High yielding, semidwarf varieties of rice and wheat deplete soil nutrients at a much higher rate than using the traditional low yielding varieties. Depletion is much greater in the rice-wheat rotation because two high yielding crops are grown per year. The system has become unsustainable in that soil nutrients in the form of manure are not being recycled into the system but end up in the atmosphere as smoke. Farmers are trying to compensate by using ever increasing amounts of inorganic fertilizer to achieve yields they were accustomed to when the rice-wheat rotation was adopted.

The increased rates of nitrogen have spawned secondary pest problems on rice including bacterial leaf blight, brown and whitebacked planthoppers, and leaffolder. Outbreaks of these pests occur annually throughout the region where higher rates of fertilizer are used (Majid et al. 1979, FAO 1986). These rates are higher than optimal for economic yields and tolerance of pest damage.

Two techniques—diagnostic survey and systems analysis—were used to rapidly describe the current rice-wheat farming systems in the target countries, determine the interrelationships of each subsystem, and pinpoint key areas for research. A diagnostic survey is a good interdisciplinary tool to quickly describe the farming system, determine constraints to higher production, and identify research priorities (Fujisaka 1990, 1991b). The tech-

74

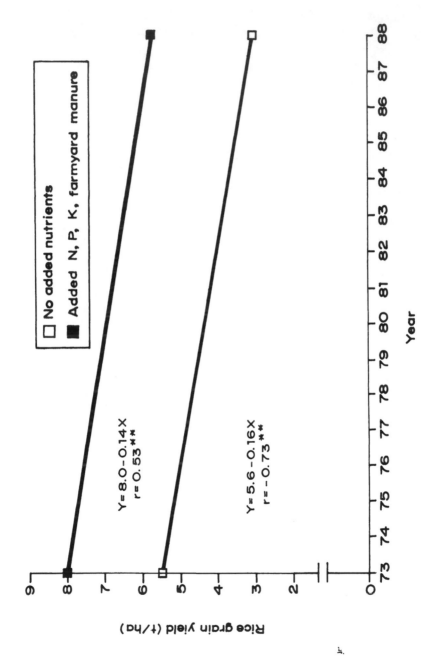

FIGURE 3.5 Trends in rice yield for a rice-wheat rotation with and without added nutrients. Pantnagar University, Uttar Pradesh, India, 1973-1988.

nique of systems analysis with flow diagrams (Conway, 1986, Norton et al. 1991) was used during the diagnostic survey to show the components and interrelationships of a subsystem such as the rice-wheat cropping pattern.

The team was addressing the larger problem of identifying possible reasons for yield stagnation and one possibility was the buildup of pests. A flow diagram of the problem in the Tarai belt of Uttar Pradesh was constructed by an interdisciplinary team to show causes and effects in the pest problem diagnosis (as shown in Figure 3.6). Pest problems were believed to build up from the extended cropping period of the rice-wheat system for those pests common to both crops. High rates of inorganic nitrogen had a double-edged effect of increasing pest number and increasing the crops' ability to tolerate pest damage.

Minimum tillage is being introduced in the region as a method of planting the wheat crop soon after rice harvest under better moisture and temperature conditions. Longer delays in wheat establishment result in progressively lower yield potential from cold and moisture stress (Inayatulla et al. 1989). Higher populations of stem borers which overwinter in rice stubble may now occur because the stubble is left standing in minimum tillage fields (Rehman and Salim 1990). Inayatulla et al. (1989) agreed that minimum tillage resulted in higher stem borer numbers but mortality was sufficiently high overwinter so that damage to the following rice crop was no worse than under the high tillage system when rice stubble was plowed under. More research is needed to quantify the stem borer overwinter survival in a number of locations.

Continuous rotation of rice-wheat has favored a weed *Phalaris minor* Retz. in wheat. A systems analysis based on interviews with farmers and experience of researchers was carried out by a diagnostic team; this produced a flow diagram showing the possible causes (as shown in Figure 3.7). Farmers traditionally plant sugarcane on one part of their land as an alternate to rice-wheat. *Phalaris* is suppressed in sugarcane. But not all farmers are allowed to grow sugarcane since hectarage is government-controlled. Ill drained fields also cannot be planted to sugarcane. This caused farmers to use herbicides as labor is scarce for hand weeding. Timing of herbicide application is critical and farmers, not that knowledgeable about herbicide use, only make matters worse; often *Phalaris* becomes a problem despite herbicide use.

Due to lack of sprayers, farmers mix herbicide in sand or urea and spread it by hand, creating a health risk. The research team will address new ways to control *Phalaris* as it will be a continuing problem for farmers who cannot rotate rice-wheat with sugarcane.

Upland Rice Cropping Patterns

Upland rice represents a distinctly different ecosystem and requires different pest control strategies. It has received more attention from research

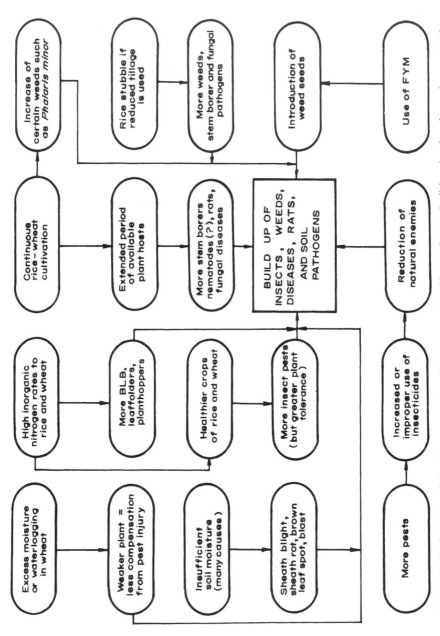

FIGURE 3.6 Causes of the long-term farming system problem of pest buildup in the rice-wheat rotation. Tarai zone of Nepal and India (Fujisaka, submitted).

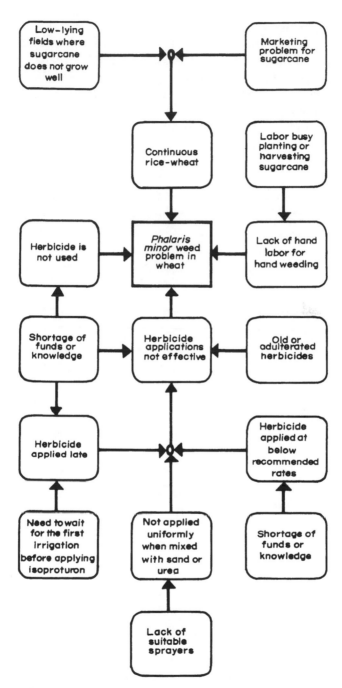

FIGURE 3.7 Causes of the *Phalaris minor* problem in wheat in the rice-wheat rotation. Tarai zone of India (Hobbs et al., 1991).

during the past several decades. As opposed to lowland rice, upland rice is often a minor crop and pest buildup is highly influenced by the composition of the surrounding flora (Litsinger et al. 1987a). Many insect pests—notably soil pests and seedling maggots—do not occur in the lowlands.

Soil insect pests tend to have wide host ranges and long life cycles. They are principally a problem in upland environments dominated by grassland fallows as flooding during a wetland rice crop kills them. Long life cycles mean, however, that controlling them in one crop may offer protection in subsequent crops. Pest management therefore can occur at the systems level.

Melolonthid white grubs, (*Holotrichia mindanaoana* Brenske and *Leucopholis irrorata* (Chevrolat) in the Philippines have either one or two year life cycles. Species are highly indigenous but their habits are similar as they feed on the roots of most plants, preferring grasses. Mole cricket, *Gryllotalpa orientalis* Burmeister as a soil pest also has a wide host range. Its life cycle is 5–6 months and the adults can live for another 6 months. Both white grubs and mole crickets cause greatest damage during the early vegetative stage of a crop. Feeding on the germinating seedlings or roots kills the plants, causing sparse stands.

In the upland tropics, two or three crops can be planted sequentially in one year depending on the length of the rainy season. Soil pests are more vulnerable to control measures during their young larval or nymphal stages. Insecticides or biological control agents (nematodes or pathogens as microbial insecticides), to be effective at minimal dosages, should be applied in the soil during land preparation in the seed furrow or incorporated during interrow cultivations (Litsinger et al. 1983). Upland farmers generally have low cash resources so a single application of material is advisable. Local production of biological control agents, however, can reduce the cost substantially (Tryon and Litsinger 1988).

Efficacy against soil pests at lowest dosage is best achieved by timing the application after the eggs hatch. In areas of monsoon Asia with distinct dry seasons, the crop year begins with the onset of the rains when farmers and insect pests resume their cropping and biological cycles, respectively. Egg laying usually peaks in relation to the rainfall pattern in the early rainy season. The dry season causes the pest population to synchronize its life cycle as normally only one insect life stage is adapted to survive the dry season. In the case of white grubs, it is the last instar larva which aestivates. Mole cricket adults emigrate from nearby wet habitats.

As rainy seasons are highly variable in their onset (in terms of frequency and intensity of rainfall), it is usually best to follow the farmers' planting schedule as he has had the most experience and he normally follows a set of indigenous rules. There will be a starting date before which the farmer will not plant, regardless of rainfall. After that starting date the farmer waits for a period of heavy rainfall to initiate sowing; this provides a degree of planting synchrony among farms in each region. Control of these pests in rice offers protection to the following crop, usually maize. In the case of white grubs with two-year life cycles, control will last for two years. In Mindanao, the majority of eggs are laid during even-numbered years. There is an insignificant percent of the population that lays eggs on odd numbered years as well.

The time of incorporation of the insecticides should be coordinated with local farmer practices. Banding the insecticide concentrates the material in seed furrows or furrows made during interrow weed control operations. If the latter time is chosen, an additional effort to cover the opened furrows is required; this may or may not be done by farmers. These trade-offs between timing and tillage methods should be tested under local conditions using various low dosages of the materials being tested.

Another upland pest, the seedling maggot *Atherigona oryzae* Malloch, attacks both rice and maize during the vegetative stage. Its control strategy also requires a thorough understanding of the farming system. The larvae feed on decaying tillers which they sever. The tillers are only susceptible when small making the crop vulnerable only the first four weeks after emergence. Rice can send out secondary and tertiary tillers, whereas maize, which does not tiller, cannot compensate for the injury. Land preparation for maize is less intensive, therefore when the rains come, farmers prepare their land for maize first; rice normally follows several weeks to a month later. Maize, being more drought tolerant, is better suited to be planted at the beginning of the rainy season when rainfall may not be frequent.

Early planting escapes seedling maggot infestation but in most years, the maize crop escapes damage while the rice crop is only moderately damaged. Other management methods to control seedling maggot on rice include sowing at higher densities, better crop nutrition, or low-dosage chemical seed treatment. The first two methods increase the crop's ability to tolerate the damage.

The planting time of rice or maize may be dictated by a number of factors. In Claveria, northern Mindanao, Philippines, farmers plant either maize-maize or a single rice crop. In a given year, the proportion of the farm planted to each depends on the current market prices of the crops, soil fertility (related to degree of erosion), and weed incidence. Based on past observations, rice is chosen for the low fertility soils and maize is selected for weedy fields (Fujisaka 1989). Rice performs relatively better in more acidic soils (more eroded) while weeds are easier to control by interrow cultivation in a maize crop.

In 1987, the rainy season began late, but in a pattern of continuous rains, both crops were planted before seedling maggot damage occurred (as shown in Figure 3.8). However, if the rains come before the starting date of

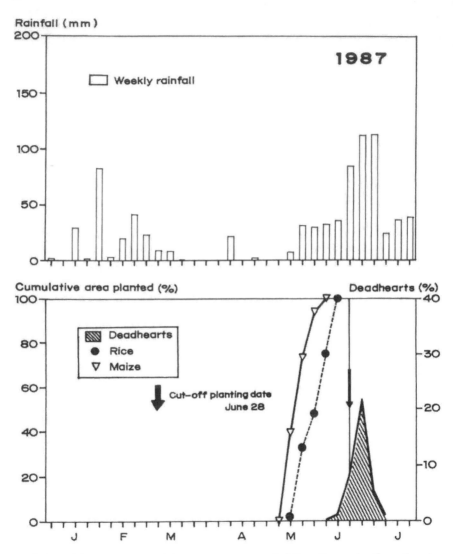

FIGURE 3.8 Relationship of the pre-monsoon rainfall pattern, planting of maize and rice, and incidence of seedling maggot *Atherigona oryzae*. Claveria, Misamis Oriental, 1987 and 1991.

FIGURE 3.8 (Continued)

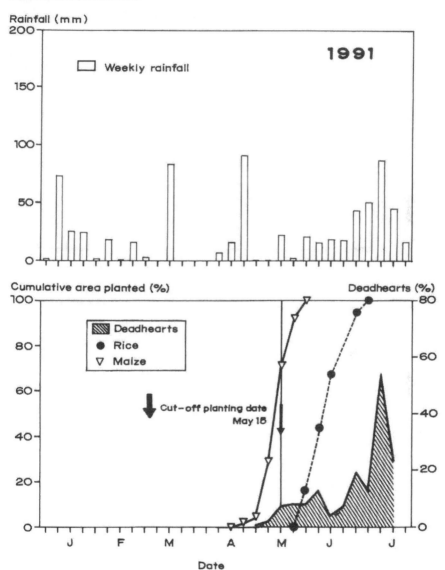

the farmer (such as in 1991 when heavy rains were followed by dry periods), the seedling maggot population builds up to attack both crops. Farmers then have to make choices to sow at higher densities, to use higher levels of fertilizer, or to apply an insecticide seed treatment. These decisions should be made before planting time.

In the uplands, there is often a strong interaction between crop management practices and the crop's ability to tolerate damage, particularly if varieties with at least moderate tillering ability are chosen. In a study in Tupi, South Cotabato, in southern Philippines, we see the interaction of seeding rate and fertilizer on nematode and insect control by nematicides and insecticides. Upland farmers who till poor soils normally seed rice at low rates and do not apply fertilizer. In more favorable areas such as Tupi, farmers seed at higher rates and apply 60–90 kg N/ha (Fujisaka 1992).

In a recent study, increasing only the seeding rate from 50 to 100 kg/ha resulted in an insignificant (0–6%) yield gain on unfertilized plots. Yields increased by 22–69% at the 50 kg/ha seeding rate when 90 kg N/ha were added. A further yield increase occurred (44–82%) when fertilizers were applied to plots at the 100 kg seed/ha sowing rate. Adding nematode control increased yields by 106–135%. Highest yields were obtained with the 100 kg/ha seeding rate, 90 kg N/ha, and nematode control. Additional insect control did not improve yield. At the Claveria site with poorer soils, insect pest damage is a factor that limits high yield; it was not so in Tupi. The basis of this study came from an observation of very low yields when nitrogen was not applied. It turned out that nitrogen was indirectly contributing to yield as it allowed the rice crop to outgrow nematode damage.

Tupi farmers have the choice of increasing their yield to reach three plateaus. The first plateau is reached by applying only fertilizer; the second, by increasing the seeding rate and applying fertilizer; and the third, by additionally controlling nematodes. Research is now under way to develop less cash-dependent nematode control methods: 1) rotation with a legume or 2) treating the seed with indigenous soil bacteria to prevent nematode establishment.

One of the yield limiting problems in the uplands is erosion caused first by overlogging and second by population pressure for land, forcing farmers to till steep slopes. Sloping land under plow cultivation quickly loses its topsoil. The reduced fertility causes yields to decline. With increasing population pressure, farmers have little chance to relocate to new lands and are faced with low production.

Contour hedgerow agriculture is a technique to stabilize sloping lands where strips of perennial vegetation are placed 5–10 m apart (closer together in steeper slopes) to hold the soil (Fujisaka 1989). In a few years, terraces form and erosion is dramatically slowed down. Farmers can make the perennial strips along the contour with simple technology using a cheaply constructed wooden A-frame.

Of interest to pest management is the choice of plant species to grow in the untilled vegetative strip which can range from naturally occurring weeds, nitrogen-fixing legumes (annual forbs or trees), or perennial grasses. As the system matures, the farmer may choose to plant cash crops in the vegetative strips to better utilize the land. The choice of vegetation depends on the needs of the farmer and the local adaptability of the species to soil and climate.

There is a succession of possible choices, each with an increasing level of management. Naturally occurring weeds or leguminous forbs are used if the farmer is only concerned with reducing erosion. Perennial grasses are used if the farmer wants to cut and carry forage to penned animals. Multipurpose nitrogen fixing trees can be grown if the farmer wants either fuelwood, soil mulch, or a biological source of nitrogen from incorporating the leaves as fertilizer (MacLean et al. 1992).

The farmers in northern Mindanao first chose to adopt the system requiring the least management. This was part of a farmer participatory project. Farmers were shown the basic idea as a result of a trip to a World Neighbors' site in Cebu, central Philippines (Fujisaka 1989). They eventually modified the contour terrace technology to suit their needs. Perennial trees were laborious to plant and care for, and when grown they competed with the annual crops planted in the alleyways.

Farmers did not adopt using the leaves of the perennial trees for mulch or organic fertilizer, again due to labor shortage. Some farmers are planting *Gmelina arborea* Roxb., a fast growing tree, in the vegetative strips. Due to deforestation, a local wood shortage has driven up the price of hardwood such as *Gmelina*.

The choice of species however can have an impact on pest populations. In Indonesia, *Brachiara brizantha* (Hochst. ex A. Rich.) Stapf (on the face of the terrace riser) and *Setaria anceps* Stapf ex Massey (on the lip of the bund along the contour) were introduced on the vegetative strips (for cut and carry goat raising) in Ciamis near Jogjakarta (AARD 1987). Within a few years, the upland rice and maize grown in the terraced alleyways were under heavy attack by white grubs that bred freely in the introduced grasses. The *Brachiara* and *Setaria* should be replaced with an equally adapted forage grass that is less conducive to white grub population buildup. On hindsight, it would have been very difficult to predict the outcome of increased white grubs from the introduction of a forage grass. Trial and error will remain the best research method for many years to come. It is important to have an interdisciplinary team on such a project in order to detect problems when they first occur so that timely changes can be made.

The establishment of strips of perennial vegetation on terraced contours creates a form of intercropping which, as the example in Indonesia showed, has an impact on insect pests and their control. Natural enemies of rice insect pests could benefit from these strips by way of food and refuge and thereby encourage greater biocontrol activity in upland systems. For various vegetative strip habitats, which natural enemy populations colonize these strips, what their impact will be, and how they should be managed remain to be discovered.

The FSR approach is holistic and the intensive site descriptive process encourages comparisons to be made between sites. Upland systems tend to be more diverse than lowland sites in terms of soils and surrounding flora. Lowland systems, with puddling, tend to homogenize soils. Farmers take advantage of the favorable water-retaining properties of lowland sites to cultivate rice, and lowland areas tend to become rice bowls—large contiguous areas devoted to rice—only broken up by narrow bunds, irrigation canals, and roads.

In sharp contrast are upland systems which have diverse soils, unmodified by rice cultivation and a botanical succession of ecosystems beginning with slash-and-burn in mostly forested areas to vast underpopulated grasslands that exist after extensive deforestation to re-forestation by a diversity of economic crops upon intensive human settlement. Pest problems dramatically change along this ecologically successional continuum.

Forest animals dominate in the slash-and-burn system where rice is often dibbled among smoldering tree stumps left after the initial clearing fire. The new settlers plant rice for their own consumption as they need a fast-growing cereal. The cleared land can be grown to rice for several years before the natural fertility gives out and grassy weeds become dominant.

In two slash-and-burn sites where I have worked—Sitiung, Sumatra in Indonesia and Magsaysay near Siniloan in Laguna, Philippines—the small upland rice plantings were under severe risk from forest animals. In Sitiung, farmers frequently lost their crop to wild pigs, monkeys, rats, birds, elephants, squirrels, and even rhinoceros (Fujisaka et al. 1991). Rice bug *Leptocorisa oratorius* (F.) was also mentioned as a severe pest, becoming concentrated on the rice crop near harvest. In Magsaysay, attempts to carry out yield loss trials over three years essentially failed, as invariably, rats destroyed each of the four farms (replicates) right before harvest. These vertebrate pests and rice bugs can seek out isolated fields and totally destroy them. The animals have been displaced by the removal of the forest and rice is an attractive food for them. The pests concentrate on the small fields, causing such devastation that farmers often give up planting rice. Unfortunately, research can provide but little technology to cope with vertebrate pest control in such low-input systems. In upland areas of Laos, under similar conditions, farmers complained about rats, wild pigs, and birds as rice pests (Fujisaka 1991c).

After clearing and before intensive human settlement, perennial grasses replace the forest. The grassland becomes the source of a number of rice pests particularly soil pests, seedling maggot, and rice bug. Claveria, northern Mindanao represents this upland rice ecosystem type and losses due to these chronic pests typically average 20–40% (Litsinger et al. 1987b).

Upon human settlement and permanent plow agriculture, the grasslands are essentially removed and replaced by annual and perennial crops. The land is plowed and harrowed repeatedly each year leading to the destruction of soil pests and seedling maggot. Tanauan, Batangas and Tupi, South Cotabato represent such a third upland rice ecosystem in the Philippines. Yield losses to pests are reduced to less than 5% (Litsinger et al. 1987b).

A comparison of pests and yield losses between upland rice sites allowed this scheme to be developed and showed a trend opposite than found in the lowlands. In lowland ecosystems, extensive and intensive rice systems spawned numerous pest problems (Litsinger 1989), whereas extensive but nonintensive upland rice systems reduced pest problems. More site comparisons are needed, however, to confirm this observation.

Conclusion

This chapter shows that the complexity inherent in tropical rice agroecosystems occurs at all levels. Integration of pest management tactics at the farming systems level should not be shied away from because of the systems' complexity per se. The challenge is the discovery of beneficial interactions (many of which may not be obvious) that can be turned to advantage in pest suppression. An interdisciplinary team applying current FSR methods can unravel the seemingly complicated system. It is hoped that the reader will be able to perceive that there can be simplicity beyond complexity: what is complex is not necessarily complicated. Examples are given to show a wide range of possibilities but not all factors in a farming system interact significantly with all others. Frustrations expressed by some (Newsom 1980, Goodell 1984, Stoner et al. 1986, Way 1987) at the intricacies of multipest interactions, control tactics, and integration of IPM at the farming system level should be taken as an opportunity to analyze them.

What was lacking in earlier efforts was the formation of truly interdisciplinary teams. IPM teams formed at universities, where disciplinary lines are often jealously guarded (Russell et al. 1982), have greater difficulties in achieving interdiscipline than those formed at departments of agriculture or international centers. The team approach works because members of other disciplines are also looking for the same interactions.

Critical to their discovery is an open minded, holistic perspective where preconceived ideas and early narrowing of the research agenda—i.e., "a small and focused approach"—may miss detection of important interactions. The approach to revealing important interrelationships, however, should follow a rational manner of forming hypotheses based on information continuously gathered at the site rather than testing an array of crop production components in a factorial design (Shri Ram and Gupta 1989). Results of systems analysis should point out specific areas for research. Together, the team should determine the research agenda to allow more opportunities for interdiscipline.

The emerging technology should be tailored to the realities of the existing farming system and to farmers' capabilities, risk aversion, and resources (Fujisaka 1990, 1991d). The technology should also be flexible, allowing farmer selection and modification rather than recommending rigid packages of practices. Farmers differ in skills, knowledge, outlook, land tenure, and resources. They should be encouraged to modify pest control technology, such as economic thresholds, to fit their specific needs (Carlson and Müeller 1991).

IPM is complex and highly sensitive to crop management. The FSR perspective, therefore, is an ideal method by which to formulate pest management strategies for Asian rice farmers. The FSR approach has been advocated by others (Altieri 1984, Matteson et al. 1984, Reichelderfer and Bottrell 1985, and Teng 1985). Integration with a farming systems perspective is not the last rung in the IPM ladder (Newsom 1980) as institutional obstacles to the implementation of IPM are still to be met (Kenmore et al. 1985).

From the examples of four rice farming systems, we saw how the FSR approach was applied to insect pest management. The complexities of the systems make agricultural development in the tropics site-specific. There are, of course, many similarities between sites of the same rice culture so the task is not as daunting as it may appear. Also, comparison between sites offers greater ecological insights as was shown in the upland systems.

The FSR approach is particularly useful in developing IPM tactics which involve changes in the farm enterprise—either in reducing new components or in effecting temporal or spatial arrangement of system components. The FSR approach is also useful if an integrated crop management strategy which aims to bolster the crop's ability to tolerate pest damage is considered. In the FSR perspective, new improvements in the system must yield more benefits other than just pest suppression.

IPM researchers can intervene at the systems level to develop methods that achieve greater pest suppression characteristics by changing components of the farming system. Any change in a cropping pattern, whether introducing a new crop or changing the planting time, may alter pest populations (Litsinger and Moody 1976). Examples of changes in the components of the farming system were rice-fish and rice-wheat systems in the lowlands and contour hedgerows in the uplands.

Pest-resistant varieties and the knowledge that insecticides are not as necessary as formerly believed have revived interest in rice-fish culture. This aquaculture-agriculture enterprise appears to be highly symbiotic. Adding fish to a ricefield is a form of natural enemy augmentation, a biological control tactic, with the added benefit of the fish having a cash value and increasing the yield of rice. The pest suppression benefits of fish should obviate much of the need to undertake corrective pest control action. Economic thresholds—levels of pest infestation—used for insecticide application decision making—will be re-evaluated due to the added risk of negative side effects on fish. Threshold levels will, no doubt, be raised, resulting in fewer insecticide applications. Rice yields should improve as a result of better crop husbandry practices the farmer now will provide because of the higher value of the new enterprise.

Systems analysis applied to rice-wheat systems point to a need to return organic matter to the soil. Green manure crops such as *Sesbania* may substitute for farmyard manure. Adoption of agroforestry practices, which will provide fuelwood, would also help. Green manures, or other soil-improving leguminous crops, appear to increase the cereal crop's ability to compensate for insect pest damage and may be the "nugget" (a keystone technology that creates benefits in more than one discipline) the team is looking for in the rice-wheat system.

Contour hedgerows in upland systems offer opportunities to change the cropping system for the benefit of pest suppression. A wrong choice of hedgerow species, however, can mean a new or worse pest problem. A farmer-to-farmer training project exposed farmers to erosion control technology. Farmers are now modifying that technology to fit their needs.

Changes in the temporal and spatial parameters of the cropping system were recommended to reduce the pest-carrying capacity of irrigated multirice systems. On-farm research by interdisciplinary teams led to an understanding of the causes of pest outbreaks after the introduction of Green Revolution varieties (Smith 1972, Reddy 1973). The holistic solution, of creating a rice-free period during the year over a scale of hundreds of hectares, was well understood by farmers (Goodell 1984). Area wide synchronous cropping schedules to create rice-free periods have been used to control tungro in Sulawesi, Indonesia (Manwan et al. 1987) and brown planthopper epidemics in Java (Oka 1983). It is also a solution for other pest and soil problems in addition to insect problems.

Limiting the cropping pattern to two rice crops per year allows more drying of the soil, a solution to problem soils, nematodes, and weeds. The results, however, are not all positive. Synchronous planting places a high demand on transplanting labor, machinery, and credit (Goodell et al. 1981). Landless laborers would have fewer season long job opportunities within the farm community. There are bound to be tradeoffs in any technology. A team approach is useful in analyzing tradeoffs and arriving at the best compromise.

In upland systems, cultural methods such as early planting will dominate control strategies as farmers have little cash to purchase pesticides. One drawback in the early planting of rice is that less time is available for farmers to perform tillage operations necessary for weed control.

New crop rotations employ diversity over time and intercroppings employ diversity over space. In FSR, the spatial and temporal framework that binds conventional research to experiment station plots is removed. The experimental unit becomes the farm community where researchers can test area wide management. Due to small farm sizes, the benefits of some recommendations can only be achieved by community wide adoption on a scale of ecological significance in the order of hundreds of hectares.

Careful monitoring of farmers' practices through surveys and field observations led to recognition of the ability of modern rices to tolerate pest damage. The team is still probing to try to rationalize why crop husbandry practices differed significantly between researchers and farmers. This questioning will form the basis for greater understanding of rice culture, and hopefully will result in better technology. From an insect pest management point of view, we know what the rice crop can yield under a blanket of insecticide sprays. As a goal, we want to approach these attainable yields.

Determining yield losses from single and multiple pests with varying crop production practices is a good method to engender teamwork among scientists from different disciplines. A method has been developed to analyze complex datasets where not only pests but also other crop stresses are taken into consideration in explaining yield losses (Savary and Zadoks 1992).

It is often said that the Green Revolution rices were more susceptible to pests. It is true that new rice varieties run the risk of having increased susceptibility to pests, particularly diseases. This is a constant risk in plant breeding. IR50, for example, was highly susceptible to blast. On the other hand, we now know that modern rices have a great capacity, due to their high tillering ability, to tolerate pest damage. Outbreaks of viruses and planthoppers were not due to new varieties but due to the development of irrigation systems. The varieties were "susceptible" only because they were photoperiod insensitive and could be grown in the dry season.

In small-scale rice agriculture, insect pest management is not an attractive technology for farmers if the message is only to use less insecticide, basing decisions on economic thresholds or other criteria. The benefits of insecticide usage on rice is simply unattractive to most subsistence rice farmers (Goodell 1984). A better way to promote IPM is to combine it with other crop management components (Fenemore and Norton 1985). A new technology will be more attractive if it solves not only a pest problem but a production problem as well. An interdisciplinary team is best suited to find these technological nuggets which add more pros than cons to the system.

The concept of integrated crop management as being more holistic should supersede IPM, particularly for cereal crops where cosmetic damage can be tolerated (Teng 1985). Examples of integrated crop management where pest management practices have been incorporated into crop husbandry have been developed to solve several problems at the same time, as a means of improving farmer adoption and reducing the need for expensive chemical control (Nissen and Elliot Juhnke 1984, Bäumgartner et al. 1990). Integration of IPM into crop husbandry practices bolsters the plants' ability to tolerate not only pest damage but other crop stresses as well.

The rice-wheat project showed how the FSR tools of diagnostic surveys and systems analysis can describe the many interdependencies in the system. Producing flow diagrams is a good method to generate interdisciplinary discussion and to determine research priorities.

Rice is not highly adapted to the drought prone uplands and yields are low as the crop is under constant risk of perturbations. Upland farmers generally have fewer resources than those living in the lowlands, making pest control more challenging. Upland environments are highly diverse and offer greater opportunities for environmental management. Interdisciplinary teamwork is particularly important in the uplands because many factors are necessary for achieving high yields. Following the law of the minimum, if only one factor is missing, yields will be low even if high levels of the other factors are applied. The path to high yields is a step by step process. Due to long life cycles of most soil pests, the economics of their control is enhanced as the benefit is spread to more than one crop.

Erosion control is a first step to pest management in the uplands as the strategy is to grow as healthy a crop as possible so that, to a large degree, damage from pests can be tolerated. There is less scope for upland rice plant breeders to develop insect resistant varieties because higher priority is given to many other breeding objectives (Arraudeau and Harahap 1986). One breeding objective, early seedling vigor, is useful for weed management, but it will also help the plant tolerate insect pest damage.

The IPM recommendations should not be rigid to allow farmers a choice between levels of intervention. Nonadoption is a sign of a lack of inappropriate technology. As we saw in an upland site, farmers preferred contour hedgerows of native grasses to the more labor intensive and competitive nitrogen fixing trees. As farmers' resources improve and their farmlands become more stable, they can then afford the greater risks of planting perennial crops. Once again, we see agricultural development as a series of small but well constructed steps each fitted into its place rather than a whole new staircase imposed on the farmers at one time.

It is stressed that the FSR approach is iterative and dynamic. The team constantly probes to understand the existing system. As an outcome, hypotheses are deduced and ideas are tested; results are analyzed only to formulate new hypotheses for further testing. Agricultural development is a stepwise process which begins and ends with the farmer.

Acknowledgments

I am indebted to my fellow team members at IRRI for providing information on the increase in pest incidence and soil problems when wetland ecosystems changed from single to double rice cropping. Information on weeds was provided by Keith Moody, on nematodes by J.C. Prot, on diseases by T.W. Mew, and on soil problems by K.G. Cassman. Ms. Nonnie P. Bunyi assisted with the typing of the manuscript.

Bibliography

Adkisson, P. L. and V. A. Dyck. 1980. "Resistant varieties in pest management systems," in F. G. Maxwell and P. R. Jennings (eds.). *Breeding plants resistant to insects*. John Wiley & Sons, N. Y. pp. 233–252.

Agency for Agricultural Research and Development (AARD). 1987. "Farming systems research. Upland agriculture and conservation. Research Highlights 1985-86." Ministry of Agriculture, Upland Agriculture and Conservation Project, Pertanian Lahan Kering dan Konservasi. Jakarta, Indonesia.

Aguda, R. M., R. C. Saxena, J. A. Litsinger, and D. W. Roberts. 1984. Inhibitory effects of insecticides on entomogenous fungi *Metarrhizium anisopliae* and *Beauveria bassiana*. *Int. Rice Res. Newsl*. 9: 16–17.

Ahmed, N. U., N. G. Hussain, and S. M. R. Karim. 1992. "On-farm rice-fish farming systems research in Bangladesh," in C. R. de la Cruz, C. Lightfoot, B. A. Costa-Pierce, and V. R. Carangal (eds.). *Rice-fish Research and Development in Asia*. ICLARM Conf. Proc. No. 24. Internat. Center Living Aquatic Res., Makati, Philippines.

Akabane, H., T. Miyazawa, M. Yamagishi, H. Yoshizawa, and T. Nishiyama. 1968. "On the phytotoxicity in rice plant caused by closed application of herbicide, DCPA, and an insecticide for the green rice leafhopper." Proc. Kanto-Tosan Plant Prot. Soc. 15: 76 (Japanese).

Allen, G.E. and J.E. Bath. 1980. The conceptual and institutional aspects of integrated pest management. *BioScience* 30: 658–664.

Altieri, M.A. 1984. Pest-management technologies for peasants: a farming systems approach. *Crop. Prot*. 3: 87–94.

Altieri, M. A. 1987. *Agroecology: the Scientific Basis of Alternative Agriculture.* Westview Press, Boulder.

Ardiwinata, R. O. 1957. Fish culture on paddy fields in Indonesia. *Proc. Indo-Pacific Fish. Coun.* 7: 119–154.

Arraudeau, M. and Z. Harahap. 1986. "Relevant upland rice breeding objectives," in *Progress in Upland Rice Research,* International Rice Research Institute, Los Baños, Philippines. pp. 189–197.

Barker, R., R. W. Herdt and B. Rose. 1985. "The Rice Economy of Asia." Resources for the Future. Washington, D.C.

Barrion, A.T. and J.A. Litsinger. 1987. Strepsipteran parasites of rice leafhoppers and planthoppers in the Philippines. *Int. Rice Res. Newsl.* 12: 37–38.

Bäumgartner, J., U. Regev, N. Rahalivavololona, B. Graf, P. Zahner and V. Delucchi. 1990. Rice production in Madagascar; regression analysis with particular reference to pest control. *Agric. Ecosystem Environment.* 30: 37–47.

Bentley, J.W. and K.L. Andrews. 1991. Pests, peasants, and publications: anthropological and entomological views of an integrated pest management program for small-scale Honduran farmers. *Hum. Organ.* 50: 113–125.

Bottenberg, H., J. A. Litsinger, A. T. Barrion, and P. E. Kenmore. 1990. Presence of tungro vectors and their natural enemies in different rice habitats in Malaysia. *Agric., Ecosyst. Environ.* 31: 1–5.

Cagauan, A.G. 1991. Fish as biological tool for aquatic weed management in integrated rice-fish culture system. *Biotrop Special Publication* 40: 217–229.

Carbonell, P. U. and B. Duff. 1980. "Pest control practices, material and labor cost of irrigated and rainfed farmers in Nueva Ecija," in *Proceedings of the 11th Annual Scientific Meeting, Baybay, Leyte.* Crop Science Society of the Philippines, Los Baños.

Carlson, G. A. and R. A. E. Müeller. 1991. "Farmers' perceptions, education and adoption of pest management strategies by small farmers," in *11th International Congress of Plant Protection.* E. D. Magallona (ed.). Manila, Philippines. pp. 147–154.

Cassman, K. G., M. J. Kropff, and Yan Zhen-De. 1992. "A conceptual framework for nitrogen management of irrigated rice in high-yield environments." *Proceedings of the 1992 International Rice Research Conference, 21–25 April 1992.* IRRI, Los Baños, Philippines.

Castillo, M.B., M.S. Alejar, and J.A. Litsinger. 1976a. Nematodes in cropping patterns. III. Composition and populations of plant parasitic nematodes in selected cropping patterns in Batangas. *Philipp. Agric.* 60: 285–292.

Castillo, M. B., M. B. Arceo, and J.A . Litsinger. 1976b. Nematodes in cropping patterns. IV. Populations of plant parasitic nematodes associated with cropping patterns under different rice-growing environments in Manaoag, Pangasinan. *Philipp. Phytopathol.* 12: 24–29.

Chambers, R., A. Pacey, and L.A. Thrupp. 1989. *Farmer First: Farmer Innovation in Agricultural Research.* Intermediate Technology Publ., London.

Chapman, G., S. Tangpoonpol, and S. Chantraniyom. 1992. "The role of the cultured fish common carp (*Cyprinus carpio*), nile tilapia (*Oreochromis niloticus*) and the silver barb (*Puntiius gonionutus*) as agents of rice insect pest and disease control, and an analysis of stomach contents, in lowland paddies of North-

East Thailand," in *Rice-fish Research and Development in Asia*. C.R. dela Cruz, C. Lightfoot, B.A. Costa-Pierce, and V.R. Carangal. (eds.). ICLARM Conf. Proc. No. 24. Intern. Center Living Aquatic Res.. Makati, Philippines.

Chaudhuri, H. 1985. "Rice-fish culture." Presented, Internat. Rice Res. Conf., 1–5 June, IRRI, Los Baños, Philippines.

Chelliah, S. and A. Subramanian. 1972. Influence of nitrogen fertilization on the infestation by the gall midge, *Pachydiplosis oryzae* (Wood-Mason) Mani in certain rice varieties. *Indian J. Entomol.* 34: 255–256.

Chien, C. C. 1980. Studies on the sheath rot disease and its relation to the sterility of rice plants. *Plant Protection Bulletin* 22: 31–39. (Chinese, Engl. Sum.).

Conway, G. R. 1986. *Agroecosystems Analysis for Research and Development*. Winrock Internat. Inst. Agric. Devel., Bangkok.

Daamen, R. A., F. G. Wijnands, and G. van der Vliet. 1989. Epidemics of diseases and pests of winter wheat at different levels of agrochemical input. A study on the possibilities for designing an integrated cropping system. *J. Phytopathol.* 125: 305–319.

Datta, S. K., D. Konar, P. K. Banerjee, S. K. De, P. K. Mukhohadhyay, and P.K. Pandit. 1986. Prospects of increasing food production in India through different systems of paddy-cum-fish culture in freshwater areas: a case study. *Int. Rice Comm. Newsl.* 35: 31–39.

De Datta, S. K. and H. C. Jereza. 1976. The use of cropping systems and land and water management to shift weed species. *Philippine J. Crop Sci.* 1: 173–178.

De Datta, S. K. 1981. *Principles and Practices of Rice Production*. John Wiley & Sons, New York.

dela Cruz, C. G., J. A. Litsinger, and F. Paragna. 1981. Tillage implements for soil incorporation of carbofuran granules in rainfed wetland fields. *IRRI Newsletter* 6: 17.

DeWalt, B. R. 1985. Farming systems research. *Hum. Organ.* 44: 106–114.

Dilts, R. 1990. IPM in action: a "crash" training programme implemented against the rice stem borer outbreak in Indonesia. *FAO Plant Prot. Bull.* 38: 89–93.

Eastop, V. F. 1981. "Wild hosts of aphid pests," in *Pests, Pathogens and Vegetation*. J.M. Thresh (ed.). Pitman, London. pp. 285–298

Estores, R. A., F. M. Laigo, and C. I. Adordionisio. 1980. "Carbofuran in rice-fish culture," in *Rice-fish Research and Development in Asia*. ICLARM Conf. Proc. No. 4. Internat. Center Living Aquatic Res. Makati, Philippines. pp. 53–57.

Fabellar, L. T. and E. A. Heinrichs. 1986. Relative toxicity of insecticides to rice planthoppers and leafhoppers and their predators. *Crop Protection* 5: 254–258.

Farrington, J. 1977. Research-based recommendations versus farmers' practices: some lessons from cotton spraying in Malawi. *Exp. Agriculture* 13: 9–15.

Fedouruk, A. and W. Leelapatra. 1985. *Rice Field Fisheries in Thailand*. Report of the Depart. Fisheries, Bangkok. 66 p.

Fenemore, P.G. and G.A. Norton. 1985. Problems of implementing improvements in pest control: a case study of apples in the UK. *Crop Protection* 4: 51–70.

Fisher, R.W. 1989. The influence of farming systems and practices on the evolution of the cotton-boll weevil agroecosystem in the Americas—a review. *Agric. Ecosyst. Environ.* 25: 315-328.

Flinn, J.C. and S.K. De Datta. 1984. Trends in irrigated rice yield under intensive cropping at Philippine research stations. *Field Crop Research* 9: 1–15.

Food and Agricultural Organization (FAO). 1986. Outbreak of pests and diseases: Bangladesh, Burma, China, India, Pakistan. *FAO Asia and Pacific Plant Prot. Comm. Newsl.* 24:30–34.

Frio, A.L. and E.C. Price. 1978. Changes in cropping systems, Cale, Tanauan, Batangas, 1973–1977. Social Sciences Division, IRRI, Los Baños, Philippines.

Fujisaka, S. 1989. A method for farmer-participatory research and technology transfer: upland soil conservation in the Philippines. *Experimental Agriculture* 25: 423–433.

Fujisaka, S. 1990. Rainfed lowland rice: building research on farmer practice and technical knowledge. *Agricul. Ecosyst. Environ.* 33: 57–74.

Fujisaka, S. 1991a. What does "build research on farmer practice" mean? Rice crop establishment (beusani). Eastern India as an illustration. *Agricultural Human Values* 8: 93–98.

Fujisaka, S. 1991b. A set of farmer-based diagnostic methods for setting post "green revolution" rice research priorities. *Agricultural Systems* 36: 191–206.

Fujisaka, S. 1991c. A diagnostic survey of shifting cultivation in northern Laos: targeting research to improve sustainability and productivity. *Agroforestry Systems* 13: 95–109.

Fujisaka, S. 1991d. Improving productivity of an upland rice and maize system: farmer cropping choices or researcher cropping pattern trapezoids. *Experimental Agriculture* 27: 253–261.

Fujisaka, S. 1992. "Taking farmer knowledge and technology seriously: seeding and weeding upland rice in the Philippines," in *Indigenous Knowledge Systems: the Cultural Dimension of Development*. M. Warren, D. Brokensha, and L.J. Slikkerveer (eds.). Keegan Paul, London.

Fujisaka, S., A. Dapusala, and E. Jayson. 1989. Hail Mary, kill the cat: a case of traditional upland crop pest control in the Philippines. *Philippine Q. Cultural Soc.* 17: 202–211.

Fujisaka, S., G. Kirk, J. A. Litsinger, K. Moody, N. Hosen, A. Yusef, F. Nurdin, T. Naim, F. Artati, A. Aziz, W. Khatib, and Yustisia. 1991. "Wild pigs, poor soils, and upland rice: a diagnostic survey of Sitiung, Sumatara, Indonesia." IRRI Res. Pap. Ser. 155. 9 p.

Fukada, M. and T. Miyake. 1978. Effect of isoprothiolane on *Nilaparvata lugens* and *Sogatella furcifera* (Hemiptera: Delphacidae) suppression of population growth. *Japan J. Appl. Entomol. Zoology* 22: 191–195.

Gliessman, S.R., R. Garcia E., and M. Amador A. 1981. The ecological basis for the application of traditional agricultural technology in the management of tropical agro-ecosystems. *Agro-Ecosystems* 7: 173–185.

Goodell, G. 1984. Challenges to international pest management research and extension in the Third World: Do we really want IPM to work? *Bull. Entomol. Soc. America* 30: 18–26.

Goodell, G., J. A. Litsinger, and P. E. Kenmore. 1981. "Evaluating integrated pest management technology through interdisciplinary research at the farmer level," in *Conference on Future Trends of Integrated Pest Management, May 30—June*

4, 1980, Bellagio, Italy. Internat. Organ. Biological Control of Noxious Animals and Plants, Paris, France. pp. 72–75.

Harwood, R. R. 1979. *Small Farm Development: Understanding and Improving Farming in the Humid Tropics*. Westview Press, Boulder.

Heinrichs, E. A. and F. G. Medrano. 1984. *Leersia hexandra*, a weed host of the rice brown planthopper, *Nilaparvata lugens* (Stål). *Crop Protection* 3: 77–85.

Heinrichs, E. A., L. T. Fabellar, R. P. Basilio, T. C. Wen, and F. Medrano. 1984. Susceptibility of rice planthoppers *Nilaparvata lugens* and *Sogatella furcifera* (Homoptera: Delphacidae) to insecticides as influenced by level of resistance in the host plant. *Environmental Entomology* 13: 455–458.

Heinrichs, E. A. and F. G. Medrano. 1985. Influence of N fertilizer on the population development of brown planthopper (BPH). *IRRI Newsletter* 10: 21–22.

Heinrichs, E. A., H. R. Rapusas, G. B. Aquino, and F. Palis. 1986. Integration of host plant resistance and insecticides in the control of *Nephotettix virescens* (Homoptera: Cicadellidae), a vector of rice tungro virus. *J. Econ. Entomol.* 79: 437–443.

Herzog, D. C. and J. E. Funderurk. 1985. "Plant resistance and cultural practice interactions with biological control," in *Biological Control in Agricultural IPM Systems*. M.A. Hoy and D.C. Herzog (eds.). Academic Press, New York. pp. 67–88

Hibino, H. 1989. Insect-borne viruses of rice. *Adv. Dis. Vector Research* 8: 209–241.

Hobbs, P. R., G. P. Hettel, R. P. Singh, Y. Singh. L. Harrington, and S. Fujisaka. 1991. "Rice-wheat cropping systems in the tarai areas of Nainital, Rampur and Pilibhit districts in Uttar Pradesh, India: diagnostic surveys of farmers' practices and problems, and needs for further research." Centro Internacional de Mejoramiento de Maiz y Trigo (CIMMYT), Mexico, D.F.

Hollis, J. P. and S. Keoboonrueng. 1984. "Nematode parasites of rice," in *Plant and Insect Nematodes*. W.R. Nickle (ed.). Marcel Dekker, New York. pp. 95–146.

Hora, S. L. and T. V. R. Pillay. 1962. *Handbook on Fish Culture in the Indo-Pacific Region*. Fish Biol. Tech. Paper 14. FAO, Rome.

Inayatullah, C., Ehsan-Ul Haq, Ata-Ul Mohsin, A. Rehman, and P. Hobbs. 1989. "Management of rice stem borers and the feasibility of adopting no-tillage in wheat." Pakistan Agricultural Research Council, Islamabad.

International Rice Research Institute (IRRI). 1973. "Entomology," in *IRRI Annual Report for 1972*. IRRI, Los Baños, Laguna, Philippines. pp. 163–188.

IRRI. 1974. "Entomology," in *IRRI Annual Report for 1973*. IRRI, Los Baños, Laguna, Philippines. pp. 209–233.

IRRI. 1987. "Weed-insect interactions," in *Annual Report for 1986*. IRRI, Los Baños, Laguna, Philippines. pp. 257–258.

IRRI. 1990. "Tropical rice-wheat systems," in *Program Report for 1989*. IRRI, Los Baños, Laguna, Philippines.

IRRI. 1992. "Tropical rice-wheat systems," in *Program Report for 1991*. IRRI, Los Baños, Laguna, Philippines.

Ishii, S. and C. Hirano. 1963. Growth responses of larvae of the rice stem borer to rice plants treated with 2,4-D. *Entomol. Exp. Appl.* 6: 257–262.

Israel, P. 1967. "Varietal resistance to rice stem borers in India," in *The Major Insect Pests of the Rice Plant*. Johns Hopkins Press, Baltimore. pp. 391–403.

Janzen, D. H. 1973. Tropical agroecosystems. *Science* 182:1 212–1219.

Kartohardjono, A. and E. A. Heinrichs. 1984. Populations of the brown planthopper, *Nilaparvata lugens* (Stål) (Homoptera: Delphacidae), and its predators on rice varieties with different levels of resistance. *Environmental Entomology* 13: 359–365.

Kashiwagi, Y. and Y. Nagai. 1975. Correlation of occurrence between rice leaf blast and rice planthoppers. *Proc. Association Plant Protection, Sikoku*. 10: 1–6.

Kenmore, P. E. 1987. "Crop loss assessment in a practical integrated pest control program for tropical Asian rice," in *Crop Loss Assessment and Pest Management*. P.S. Teng (ed.). Amer. Phytopath. Soc. Press, St. Paul, Minnesota. pp. 225–241.

Kenmore, P. E., K. L. Heong, and C. A. Putter. 1985. "Political, social and perceptual aspects of integrated pest management programmes," in *Proceedings of the Seminar on Integrated Pest Management in Malaysia*. B.S. Lee, W.H. Loke, and K.L. Heong (eds.). Malaysian Plant Prot. Soc., Kuala Lumpur. pp. 47–67

Kenmore, P. E., J. A. Litsinger, J. P. Bandong, A. C. Santiago, and M. M. Salac. 1987. "Philippine rice farmers and insecticides: thirty years of growing dependency and new options for change," in *Management of Pests and Pesticides: Farmers' Perceptions and Practices*. J. Tait and B. Napompeth (eds.) Westview Press, Boulder. pp. 98–108.

Khan, Z. R. and R. C. Saxena. 1985. Behaviour and biology of *Nephotettix virescens* (Homoptera: Cicadellidae) on tungro virus-infected rice plants: epidemiology implications. *Environmental Entomology* 14: 297–304.

Kiritani, K. 1977. Recent progress in the pest management for rice in Japan. *JARQ* 11: 41–49.

Kiritani, K. 1979. Pest management in rice. *Annual Review of Entomology* 24: 279–312.

Koesoemadinata, S. 1980. "Pesticides as a major constraint to integrated agriculture-aquaculture farming systems." in *Proceedings of the ICLARM-SEARCA Conference on Integrated Agriculture-aquaculture Farming Systems, Manila, Philippines, 6–9 Aug. 1979*. Internat. Center Living Aquatic Res. Makati, Philippines. pp. 45–51.

Kovitvadhi, K., N. Chantaraprapha, and T. Bhudhasamai. 1972. "Preliminary study on integrated control of rice gall midge," in *FAO International Rice Commission 14th Session of the Party on Rice Production and Protection, November 6–10*. Bangkok, Thailand.

Lam Y. M. 1991. "Cultural control of rice field rats," in *Rodents and Rice*. G. R. Quick (ed.). IRRI, Los Baños, Philippines. pp. 65–72.

Lange, W. H. Jr., K. H. Ingebretsen, and L. L. Davis. 1953. Rice leaf miner: severe attack controlled by water management, insecticide application. *California Agriculture* 7: 8–9.

Lee, S. C., D. M. Matias, T. W. Mew, J. S. Soriano, and E. A. Heinrichs. 1985. Relationship between planthoppers (*Nilaparvata lugens* and *Sogatella furcifera*) and rice diseases. *Korean J. Plant Protection* 24: 65–70.

Li, K. 1988. Rice-fish culture in China: a review. *Aquaculture* 71: 173–186.

Liao, K. T. 1980. *Rice Paddy Fish Culture.* Pearl River Fisheries Inst., Bur. Aquatic Products. Guandong Science and Technology Publ. Co., Guandong, China.

Lightfoot, C., A. van Dam, and C. Costa-Pierce. 1992. "What's happening to the rice yields in rice fish systems?," in *Rice-fish Research and Development in Asia.* C.R. dela Cruz, C. Lightfoot, B.A. Costa-Pierce, and V.R. Carangal. (eds.). ICLARM Conf. Proc. No. 24. Internat. Center Living Aquatic Res. Makati, Philippines.

Lim, G. S. 1970. Some aspects of the conservation of natural enemies of rice stem borers and the feasibility of harmonizing chemical and biological control of these pests in Malaysia. *Mushi* 43:127–135.

Litsinger, J.A. 1989. Second generation insect pest problems on high yielding rices. *Tropical Pest Management* 35: 235–242.

Litsinger, J. A. 1991. "Crop loss assessment in rice," in *Rice Insects: Management Strategies.* E.A. Heinrichs and T.A. Miller (eds.). Springer-Verlag, New York. pp. 1-65.

Litsinger, J. A. 1992. "Cultural, mechanical, and physical control," in *Insect Pests of Rice.* E.A. Heinrichs (ed.). John Wiley Eastern, New Delhi, India. (in press)

Litsinger, J. A. and K. Moody. 1976. "Integrated pest management of multiple cropping systems," in *Multiple Cropping.* R.I. Papendick, P.A. Sanchez, and G.B. Triplet (eds.). Special Publ. No. 27. Amer. Soc. Agron., Madison. pp. 293–316.

Litsinger, J.A., E.C. Price, and R.T. Herrera. 1980. Small farmer pest control practices for rainfed rice, corn, and grain legumes in three Philippine provinces. *Philippine Entomologist* 4: 65–86.

Litsinger, J.A., B.L. Canapi, and A.L. Alviola. 1982. Farmer perception and control of rice pests in Solana, Cagayan Valley, a pre-green revolution area of the Philippines. *Philippine Entomologist* 5: 373–383.

Litsinger, J.A., R.F. Apostol, and M.B. Obusan. 1983. White grub, *Leucopholis irrorata* (Coleoptera: Scarabaeidae): pest status, population dynamics, and chemical control in a rice-maize cropping pattern in the Philippines. *J. Econ. Ent.* 76: 1133–1138.

Litsinger, J. A. and Ruhendi. 1984. Rice stubble and straw mulch suppression of preflowering insect pests of cowpeas sown after puddled rice. *Environmental Entomology* 13: 509–514.

Litsinger, J. A., A. T. Barrion, and D. Soekarna. 1987a. "Upland rice insect pests: their ecology, importance, and control." IRRI Research Paper Series 123. 41 p.

Litsinger, J. A., B. L. Canapi, J. P. Bandong, C. G. dela Cruz, R. F. Apostol, P. C. Pantua, M. D. Lumaban, A. L. Alviola III, F. Raymundo, E. M. Libetario, M. E. Loevinsohn, and R. C.Joshi. 1987b. Rice crop loss from insect pests in wetland and dryland environments of Asia with emphasis on the Philippines. *Insect Sci. Appl.* 8: 677–692.

Loevinsohn, M. E. 1991. "Asynchrony of cultivation and the ecology of rice pests," in *International Plant Protection: Focus on the Developing World.* E.D. Magallona (ed.). Proc. 11th Internat. Congress of Plant Prot., 5–9 Oct 1987, Manila, Philippines. pp. 44–47.

Loevinsohn, M. E., J. A. Litsinger, and E. A. Heinrichs. 1988. "Rice insect pests and agricultural change." in *The Entomology of Indigenous and Naturalized Systems in Agriculture*. M.K. Harris and C.E. Rogers (eds.). Westview Press, Boulder, CO. pp. 161–182.

MacKay, K. T., G. Chapman, J. Sollows, and N. Thongpan. 1988. "Rice-fish culture in Northeast Thailand: stability and sustainability," in *Global Perspectives on Agroecology and Sustainable Agricultural Systems*. P. Allen and D. van Dusen (eds.) Agroecology Program, University of California, Santa Cruz. pp. 355–370.

Macklin, M. C. and M. V. Rao. 1991. *Rice and Wheat Production and Sustainability of the Irrigated Rice: Wheat Cropping System in India*. World Bank, New York.

MacLean, R. H., J. A. Litsinger, K. Moody, and A. Watson. 1992. The benefit of alley cropping upland rice and maize with *Gliricidia sepium* and *Cassia spectabilis* in an acid soil in Southern Philippines. *Agrofor. Systems* (in press).

MacQuillan, M.J. 1974. Influence of crop husbandry on rice planthoppers (Hemiptera: Delphacidae) in the Solomon Islands. *Agro-Ecosystems* 1: 339–358.

Majid, A., M. A. Makdoomi, and I. A. Dar. 1979. Occurrence and control of the whitebacked planthopper in the Punjab of Pakistan. *IRRI Newsletter* 4: 17.

Manwan, I., S. Sama and S. A. Rizvi. 1987. "Management strategy to control rice tungro in Indonesia," in *Proceedings of the Workshop on Rice Tungro Virus, 24-27 Sept. 1986*. Maros Research Inst. Food Crops, Minist. Agric., Sulawesi, Indonesia. pp. 92–97.

Matteson, P.C., Altieri, M. A., and Gagne, W. C. 1984. Modification of small farmer practices for better pest management. *Ann. Rev. Entomol.* 29: 383–402.

Mochida, O. (1991) Spread of freshwater *Pomacea* snails (Pilidae, Mollusca) from Argentina to Asia. *Micronesia Suppl.* 3: 51–62.

Mochida, O., H. T. Guevarra, J. A. Litsinger, and R. P. Basilio. 1991. "Golden apple snail *Pomacea canaliculata*: an introduced pest of rice," in *International Plant Protection: Focus on the Developing World*. E. D. Magallona (ed.). Proc. 11th Internat. Congress of Plant Prot., Oct. 5–9, 1987. Manila, Philippines. pp. 86–89.

Mohiuddin, M. S., Y. P. Rao, S. K. Mohan, and J. P. Verma. 1976. Role of *Leptocorisa acuta* Thun. in the spread of bacterial blight of rice. *Curr. Sci.* 45: 426–427.

Moody, K. 1990. "Pest interactions in rice in the Philippines," in *Pest Management in Rice*. B.T. Grayson, M.E. Green, and L.G. Copping. (eds.). Society of Chemical Industry, London. pp. 269–299.

Moody, K. 1991. "Weed management in rice," in *Handbook of Pest Management in Agriculture*. D. Pimentel (ed.). CRC Press, Boca Raton. pp. 301-328.

Moody, K. 1992. "Fish-crustacean-weed interactions," in *Rice-fish Research and Development in Asia*. C. R. dela Cruz, C. Lightfoot, B. A. Costa-Pierce, and V. R. Carangal. (eds.). ICLARM Conf. Proc. No. 24. Internat. Center Living Aquatic Res. Makati, Philippines.

Moulton, T.P. 1973. More rice and less fish—some problems of the "Green Revolution". *Australian Natural History* 17: 322–327.

Murty, V.S.T., K. C. Agrawal, and R. K. Gupta. 1980. Association of stem rot disease with brown planthopper infested rice. *Oryza* 17: 241.

Myint, M. M., H. R. Rapusas and E. A. Heinrichs. 1986. Integration of varietal resistance and predation for the management of *Nephotettix virescens* (Homoptera: Cicadellidae) populations on rice. *Crop Protection* 5: 259–265.

Nakasuji, F. and V. A. Dyck. 1984. Evaluation of the role of *Microvelia douglasi atrolineata* (Bergroth) (Heteroptera: (Veliidae) as predator of the brown planthopper *Nilaparvata lugens* (Stål) (Homoptera: Delphacidae). *Researches Population Ecology* 26: 134–149.

Natarajan, N. and P. C. Sundara Babu. 1988. Culture of the sorghum earhead bug, *Calocoris angustus* Lethierry in the laboratory. *Tropical Pest Management* 34: 356–357.

Newsom, L. D. 1980. The next rung up the integrated pest management ladder. *Bull. Entomol. Soc. Amer.* 26: 369–374.

Nie Dashu, Chen Yinghong, and Wang Jiangguo. 1992. "Mutualism of rice and fish in the paddy field," in *Rice-fish Research and Development in Asia.* C.R. dela Cruz, C. Lightfoot, B. A. Costa-Pierce, and V. R. Carangal. (eds.). ICLARM Conf. Proc. No. 24. Internat. Center for Living Aquatic Res. Makati, Philippines.

Nissen, S. J. and M. E. Elliott Juhnke. 1990. Integrated crop management for dryland small grain production in Montana. *Plant Dis.* 68: 748–752

Norton, G. A., J. Holt, K. L. Heong, J. Cheng, and D. R. Wareing. 1991. "Systems analysis and rice pest management," in *Rice Insects: Management Strategies.* E.A. Heinrichs and T.A. Miller (eds.). Springer-Verlag, New York. pp. 287–321.

Nwanze, K. F. and R.A.E. Mueller. 1989. "Management options for sorghum stem borers for farmers in the semi-arid tropics," in *International Workshop on Sorghum Stem Borers, 17-20 Nov 1987, ICRISAT.* Internat. Crops Res. Inst. Semi-Arid Tropics, Patancheru, A.P. India. pp. 105–116.

Oka, I. N. 1983. The potential for the integration of plant resistance, agronomic, biological, physical/mechanical techniques, and pesticides for pest control in farming system. *Indonesian Agricul. Res. Develop. Journal* 5 (1/2): 8–17.

Oka, I, N., I. Manwan. 1978. "Integrated control of the brown planthopper in Indonesia," in *The brown planthopper (*Nilaparvata lugens*).* Proc. Symp. on the brown planthopper, Pacific Science Association, Bali, Indonesia. pp. 65–77.

Ordish, G. 1966. F.A.O. symposium on integrated pest control. *PANS* 12: 35–39.

Ozaki, K. 1959. On the difference in the resistance to parathion or methyl parathion of the hibernated rice stem borer reared on different varieties of rice plant. *Botyu-Kagaku* 24: 118-123. (Japanese, Eng. Sum.).

Padmaja Rao, S. 1986. Studies on planting time on productivity of BPH resistant rice varieties. *Madras Agricultural Journal* 73: 384–388.

Palis, F. V. 1983. "Role of integrated pest management on the natural enemy population fluctuation in a lowland ricefield." MS thesis, Entomol. Depart., Univ. Philippines, Los Baños.

Perrin, R. M. 1977. Pest management in multiple cropping systems. *Agro-Ecosystems* 3: 1–16.

Philippine Ministry of Agriculture. 1978. "Masagana 99 rice culture 16 steps: irrigated transplanted." Philippine Gov't., Agricul. Info. Div., Quezon City.

Pollet, A. 1978. The pests on rice in Ivory Coast. V. Interrelationship between

Maliarpha separatella and *Pyricularia oryzae. Zeit. Ang. Entomol.* 85: 324–327 (French, Engl. abstract).

Prakasa Rao, P. S., P. K. Das, and G. Padhi. 1975. Note on the compatibility of DD-136 (*Neoaplectana dutkyi*), an insect parasitic nematode with some insecticides and fertilizers. *Indian J. Agricul. Sci.* 45: 275–277.

Prot, J. C., I.R.S. Soriano, D. M. Matias, and S. Savary. 1992. Use of green manure crops in control of *Hirschmanniella mucronata* and *H. oryzae* in irrigated rice. *Journal of Nematology* 24. (in press).

Rao, J. and A. Prakash. 1985. *Tyrophagus palmarum* (Oudemans) mite in rice seedlings and leaf sheaths. *IRRI Newsletter* 10: 13–14.

Reddy, D. B. 1973. High yielding varieties and special plant protection problems with particular reference to tungro virus of rice. *Internat. Rice Commission Newsletter* 22: 34–42.

Rehman, A. and M. Salim. 1990. Survival of rice stem borer (SB) in different cropping systems in Sindh. *IRRI Newsletter* 15: 28–29.

Reichelderfer, K. H. and Bottrell, D. G. (1985). Evaluating the economic and sociological implications of agricultural pests and their control. *Crop Protection* 4: 281–297.

Reissig, W. H., E. A. Heinrichs, and S. L. Valencia. 1982. Insecticide-induced resurgence of the brown planthopper, *Nilaparvata lugens,* in rice varieties with different levels of resistance. *Environmental Entomology* 11: 165–168.

Reissig, W. H., E. A. Heinrichs, J. A. Litsinger, K. Moody, L. Fiedler, T. W. Mew, and A. T. Barrion. 1986. *Illustrated Guide to Integrated Pest Management in Rice in Tropical Asia.* IRRI, Baños, Philippines. 411 p.

Rothschild, G.H.L. 1971. The biology and ecology of rice-stem borers in Sarawak (Malaysian Borneo). *Journal Applied Ecology* 8: 287–322.

Ruesink, W.G. 1976. Status of the systems approach to pest management. *Annual Review Entomology* 21: 27–44.

Russell, M.G., R.J. Sauer, and J.M. Barnes (eds.). 1982. Enabling interdisciplinary research : perspectives from agriculture, forestry, and home economics. *Agricul. Exper. Station, University of Minnesota, Misc. Publ.* 19 : 1–183.

Sain, M. 1988. "Studies on varietal resistance and bionomics of rice gall midge, *Orseolia oryzae* (Woood–Mason)." Ph.D. thesis, Depart. Zool., Fac. Sci. Osmania Univ., India.

Saint, W. S. and W. Coward Jr. 1977. Agriculture and behavioral science: emerging orientations. *Science* 197: 733–737.

Santa, H. 1965. Disappearance of virus transmissibility of viruliferous green rice leafhopper (yellow dwarf virus) parasitized by a pipunculid. *Proc. Kanto–Tosan Plant Prot. Soc.* 12: 66 (Japanese).

Sarbini, G. and A. Leme. 1987. "Effect of soil infested by the root nematode *Hirschmanniella oryzae* on tungro disease of rice," in *Proc. Workshop on Rice Tungro Virus, 24–27 Sept. 1986.* Maros Research Institute for Food Crops, Ministry of Agriculture, Sulawesi, Indonesia. pp. 18–20.

Saroja, R., R. Jagannathan, and N. Raju. 1987. Effect of N nutrition and rice variety on leaffolder (LF), yellow stem borer (YSB), and grain yield. *IRRI Newsletter* 12: 11–12.

Savary, S. and J. C. Zadoks. 1992. Analysis of crop losses in the multiple pathosystem groundnut rust-leafspot. III. Correspondence analysis. *Crop Protection* 11. (in press).

Shepard, B. M. and G. S. Arida. 1986. Parasitism and predation of yellow stem borer, *Scirpophaga incertulas* (Walker) (Lepidoptera: Pyralidae) eggs in transplanted and direct-seeded rice. *J. Entomological Sciences* 21: 26–32.

Shepard, B. M., H. D. Justo Jr., E. G. Rubia, and D. B. Estaño. 1990. Response of the rice plant to damage by the rice whorl maggot *Hydrellia philippina* Ferino (Diptera: Ephyridae). *J. Plant Prot. Trop.* 7: 173–177.

Shepard, B. M., M. Parducho, and G. S. Arida. 1988. Effect of flooding on black bug *Scotinophara coarctata* (F.)) egg parasitization. *IRRI Newsletter* 13: 22–23.

Shri Ram and M. P. Gupta. 1989. Integrated pest management in fodder cowpea (*Vigna unguiculata* L. Walp) in India and its economics. *Tropical Pest Management* 35: 348–351.

Singh, K. G. 1979. Resistance of rice blast disease (*Pyricularia oryzae* Cav.) in penyakit merah virus infected plants. *Malaysian Agricul. J.* 52: 65–67.

Smith, J., J. A. Litsinger, J. P. Bandong, M. D. Lumaban, and C. G. dela Cruz. 1989. Economic thresholds for insecticide application to rice: profitability and risk analysis to Filipino farmers. *J. Plant Protection Tropical* 6: 67–87.

Smith, R. F. 1972. The impact of the green revolution on plant protection in tropical and subtropical areas. *Bull. Entomol. Society America* 18: 7–14.

Smith, R. J. and N. P. Tugwell. 1975. Propanil-carbofuran interactions in rice. *Weed Science* 23 : 176–178.

Spiller, G. 1986. *Environmental Aspects of Rice-fish Production in Asia.* FAO/ RAPA Manuscript, Bangkok. 106 p.

Stoner, K. A, A. J. Sawyer, and A. M. Shelton 1986. Constraints to the implementation of IPM programs in the U.S.A. : a course outline. *Agriculture, Ecosystems and Environment* 17: 253–268.

Strong, D. R., Jr. 1979. Biogeographic dynamics of insect-host plant communities. *Annual Review of Entomology* 24: 89–119.

Teng, P. S. 1985. Integrating crop and pest management: the need for comprehensive management of yield constraints in cropping systems. *J. Plant Protection Tropical* 2: 15–26.

Tryon, E. H. and J. A. Litsinger. 1988. "Feasibility of using locally produced *Bacillus thuringiensis* to control tropical insect pests," in *Pesticide Management and Integrated Management in Southeast Asia.* P.S. Teng and K.L. Heong (eds.). Consortium Internat. Crop Prot., College Park, Maryland. pp. 73–81.

van den Berg, H., B. M. Shepard, J. A. Litsinger, and P.C. Pantua. 1988. Impact of predators and parasitoids on the eggs of *Rivula atimeta, Naranga aenescens* (Lepidoptera : Noctuidae) and *Hydrellia philippina* (Diptera: Ephydridae) in rice. *Journal Plant Protection Tropical* 5: 103–108.

van Emden, H. F. 1977. "Insect pest management in multiple cropping systems—a strategy," in *Symposium on cropping systems research and development for the Asian rice farmer.* IRRI, Los Baños, Philippines. pp. 325–343.

van Emden, H. F. 1982. "Principles of implementation of IPM," in *Proceedings of Australasian Workshop on Development and Implementation of IPM.* P. J. Cameron, C. H. Wearing and W. M. Kain (eds.) Government Printer, Auckland. pp. 9–17.

von Arx, R., P. T. Ewell, J. Goueder, M. Essamet, M. Cheikh, and A. Ben Temine. 1988. "Management of potato tuber moth by Tunisian farmers: a report of on-farm monitoring and a socioeconomic survey." International Potato Center, Lima, Peru.

Way, M. J. 1987. "Integrated pest control strategies in food production and their bearing on disease vectors in agricultural lands," in *Effects of Agricultural Development on Diseases*. Food and Agriculture Organization, Rome. pp. 107–115.

Way, M. O., A. A. Grigarick, and S. E. Mahr. 1983. Effects of rice plant density, rice water weevil (Coleoptera: Curculionidae) damage to rice, and aquatic weeds on aster leafhopper (Homoptera: Cicadellidae) density. *Environ. Entomol.* 12: 949–952.

Way, M. O., A. A. Grigarick, and S. E. Mahr. 1984. The aster leafhopper (Homoptera: Cicadellidae) in California rice: herbicide treatment affects population density and induced infestations reduce grain yield. *J. Econ. Entomol.* 77: 936–942.

Way, M. O., A. A. Grigarick, J. A. Litsinger, F. Palis, and P. L. Pingali. 1991. "Economic thresholds and injury levels for insect pests of rice," in *Rice Insects: Management Strategies*. E.A. Heinrichs and T.A. Miller (eds.). Springer-Verlag, New York. pp. 67–105.

West Africa Rice Development Association (WARDA). 1978. "Entomology," in *WARDA Research Department 1978 Annual Report*. Monrovia, Liberia. pp. 109–112.

Willis, G. D. and J. E. Street. 1988. Propanil plus methyl parathion on rice (*Oryza sativa*). *Weed Science* 36: 335–339.

Xiao Qing Xian. 1992. "Role of fish in pest control in rice farming," in *Rice-fish Research and Development in Asia*. C. R. dela Cruz, C. Lightfoot, B. A. Costa-Pierce, and V. R. Carangal. (eds.). ICLARM Conf. Proc. No. 24. Internat. Center Living Aquatic Res. Makati, Philippines.

Xu, Yuchang and Guo Yi Xian. 1992. "Rice-fish farming systems research in China," in *Rice-fish Research and Development in Asia*. C.R. dela Cruz, C. Lightfoot, B.A. Costa-Pierce, and V.R. Carangal. (eds.). ICLARM Conf. Proc. No. 24. Internat. Center Living Aquatic Res. Makati, Philippines.

Yasumatsu, K., T. Wongsiri, C. Tirawat, N. Wongsiri, and A. Lewvanich. 1980. "Contributions to the development of integrated rice pest control in Thailand." Depart. Agri., Bangkok, Thailand and Japanese Internat. Center for Agriculture, Tsukuba, Japan.

Yusope, M. 1920. Some insect pests of padi. Agric. Bull. F.M.S., Kuala Lumpur 8:187–189.

Zandstra, H. G., E. C. Price, J. A. Litsinger, and R. A. Morris. 1981. "A methodology for on-farm cropping systems research." *IRRI*. Los Baños, Philippines.

4

Implementation of IPM for Small-Scale Cassava Farmers

A. R. Braun, A. C. Bellotti, and J. C. Lozano

Introduction

Small-scale farmers grow most of the world's cassava crop over a broad range of tropical environments, but often on fragile or poor soils, and under rainfed conditions. Cassava is often one of many components in complex traditional farming systems evolved in response to risky production conditions. As the most important root crop and the fourth most important human calorie source in the tropics (Cock 1985, CIAT 1991), cassava is important in the agriculture of difficult areas where resources are limited. In contrast, the industrialized agriculture of temperate zones and the green revolution agriculture of the tropics are typified by simple farming systems, low environmental diversity, high use of purchased inputs, and moderate risk (Chambers et al. 1989). Research has benefitted industrialized and green revolution agriculture more than resource-poor farmers, who have been slower, or unable to adopt improved technology (Chambers et al. 1989). The major challenge facing tropical agricultural research and development is to improve the welfare of peasant farmers while guaranteeing the sustainability of agriculture and the conservation of the natural resource base.

In this chapter we outline a conceptual framework for implementation of ecologically sound crop protection for cassava. Crop protection develops in steps with the ultimate goal of integration at the farm level. The commodity orientation and the development of management tactics on a pest-by-pest basis are stages in this process. After characterizing successful crop protection programs, we present two case studies on the management of single major biotic constraints in cassava. We conclude with guidelines for achieving ecologically based crop protection in cassava growing areas with multiple pest and disease constraints.

Characteristics of Successful Crop Protection Programs

Ecologically sound crop protection can be based on remedial strategies with the immediate goal of reducing pesticide use, or on combinations of host plant resistance, cultural, biological and behavioral control, where pesticides are not used. Scientists developed the integrated pest management (IPM) concept in response to crises caused by excessive use of chemical pesticides (Greathead 1991). IPM is generally implemented when overdependence on pesticides leads to more frequent applications, higher concentrations, the development of resistance by pests, transformation of minor pests into serious primary pests, or when pesticide use has increased production costs to unacceptably high levels (Metcalf and Luckmann 1975). A decision by communities or societies to ban pesticides or reduce or eliminate subsidies for them, as occurred in Indonesia (Kenmore 1988, Barbier 1989), can also stimulate implementation of IPM. Once pesticide use has been reduced, ecologically-oriented, second-stage pest management practices (Prokopy et al. 1990) should be implemented with the goal of minimizing or eventually phasing out pesticide use.

Increases in pesticide use have been documented for some IPM programs (Napit et al. 1988, Hatcher et al. 1984), raising questions about the impact of IPM on environmental quality. Implementation of environmentally sound crop protection requires IPM programs that minimize, end, or prevent pesticide use

IPM remains elusive at the operational level, particularly in developing countries (NRI 1991). In 1989 The Consultative Group for International Agricultural Research commissioned a task force to identify major constraints to implementation of IPM in the tropics, and to make recommendations for overcoming them. The key constraints are thought to be 1) the biological complexity of agricultural systems, 2) lack of information feedback mechanisms that allow IPM systems to evolve with experience and, 3) inadequate decision-making skills of farmers.

IPM is information based. Ecological pest control requires detailed understanding of particular systems. Our inadequate understanding of how systems work—not biological complexity *per se*—and the high cost of research constrain the development of IPM.

Information feedback in IPM requires increased contact and communication among farmers, extensionists, and researchers so that the knowledge and experiences of farmers, the users of technology, can feed into the research agenda. Likewise, researchers and extensionists can fill gaps in farmers' understanding of biological phenomena, refine and improve farmer experimentation (Bentley 1990, Bentley and Andrews 1991) and upgrade farmers' decision-making skills. Recognizing the limitations and strengths of both scientists and farmers, this approach combines the theoretical insights and technical power of western science with indigenous knowledge of agricultural practices (Pimbert 1991).

The IPM task force identified several common characteristics of successful IPM programs (NRI 1991): 1) a high level of farmer involvement in the development of technology; 2) farmers and researchers provide feedback to each other so that the technology evolves with experience; 3) a selection of technology components provides the farmer with flexibility in incorporating improved technology into current practices; 4) interactions between different components of the pest complex and with other aspects of crop management are considered, and there is integration at the cropping system level in the development and testing of technology.

Two Case Studies in Cassava

Cassava Hornworm

In Latin America, pesticide use in cassava is low relative to other crops such as sugarcane, cotton, coffee, soybeans, wheat and citrus (Bellotti et al. 1990). But some cassava pests are more likely to incite pesticide use than others. Cassava hornworm (*Erinnys ello* (L.), Lepidoptera: Sphingidae) attacks are sporadic and unpredictable, and severe infestations cause complete plant defoliation, losses in bulk root production, and lower root quality. Farmers often react with excessive, ill-timed applications of pesticides which may cause hornworm resurgence (Bellotti and Schoonhoven 1978).

Hornworm larvae feed on cassava leaves of all ages, and on growing points and leaf buds. In simulated damage studies, yield losses in fertile soils ranged from 0 to 25% after one attack, and up to 47% after two consecutive attacks. On less fertile soils, two consecutive attacks caused losses as high as 64% (Arias and Bellotti 1984). Repeated attacks are most common when poorly timed pesticide applications fail to destroy fifth instar larvae or prepupae, but kill natural enemies that build up during the initial hornworm outbreak.

Although outbreaks are sporadic, they often occur in the rainy season when foliage is abundant. Vigorous plants can support more than 100 larvae/plant. On infertile soils, 13 fifth instar larvae can defoliate a three month old plant in 3 days (Arias and Bellotti 1984). Economic injury levels vary with plant age at the time of the attack. In cassava harvested at one year, 4.4 larvae/plant caused complete defoliation of one month-old plants and a 23% reduction in fresh root yield; at 7 months 26 larvae/plant caused an equivalent yield reduction (Arias and Bellotti 1984).

The hornworm has a large complex of natural enemies including predators, egg, larval and pupal parasites and pathogens. More than 30 species have been identified, however, they are not effective in maintaining the hornworm below economic injury levels (Bellotti et al. 1992). Although egg parasitism by *Trichogramma* and *Telenomus* spp. reaches an estimated 50% in some regions, this is not high enough to control population explosions of *E. ello* (Bellotti et al. 1992).

E. ello is a migratory species (Winder and Abreu 1976; Janzen 1986, 1987). Its strong flight abilities (Wolda 1979), broad climatic adaptation and wide host range (Janzen 1985) probably account for its distribution throughout the Neotropics. Adults migrate long distances *en masse* and the larvae feed on natural vegetation and on cultivated species including papaya, tomato, tobacco and cotton (Winder 1976). Some researchers believe the migration of the cassava hornworm evolved as a mechanism to survive low food availability, unfavorable environmental conditions and attack by natural enemies (Wallner 1987, CIAT 1991). The migration of hornworm adults reduces the effectiveness of natural enemies. In the absence of the hornworm, its natural enemies occur at low levels. When large numbers of migrating hornworm adults invade a cassava field, they lay up to 600 eggs/plant (Bellotti et al. 1992). This population explosion overcomes the capacity of natural enemies to drive hornworm numbers below economically damaging levels. The availability of foliage determines the persistence of hornworms in cassava fields, and adults migrate when the food supply falls. Natural enemy numbers then return to low equilibrium levels determined by the availability of other host or prey species.

Successful control of the hornworm requires monitoring of field populations for detection of immigrant adults or larvae in the early instars. This can be accomplished with black lights which trap flying adults, or by scouting fields for the presence of eggs and larvae (Bellotti and Reyes 1989). Once an invasion has been detected, control depends on the availability of an effective natural enemy for inundative release. Releases of *Trichogramma* spp. to augment natural levels of parasitism can be successful, however *Trichogramma* spp. prefer to parasitize recently laid hornworm eggs (Bellotti et al. 1992). The degree of synchronization required for effective release of the parasites, and the availability of sufficient numbers of *Trichogramma* spp. limit their use. In the absence of a reliable commercial source of *Trichogramma* spp., the cost of maintaining this parasite in continuous culture to guarantee availability when an *E. ello* invasion occurs is generally prohibitive for resource-limited cassava farmers.

The complexities of inundative releases of *Trichogramma* spp. suggest that a cheap, storable biological pesticide is needed. We discovered a granulosis virus of the family Baculoviridae attacking *E. ello* larvae in Colombia in the early 1970s. To study pathogenicity, prepared virus material from infected larvae was collected from the field. Larvae were then liquefied in a blender and filtered though cheesecloth. After dilution with water, the liquid was applied to cassava fields with a backpack sprayer, and first, second and third instar larvae were released in the treated fields. After 24 hours, collected larvae were placed on clean detached cassava leaves in the laboratory. After 72 hours larval mortality reached 100% (Bellotti et al. 1990). After 72 hours, 0.9 ml virus suspension/l water caused 90% mortality of all instars (as shown in Figure 4.1), however lower concentrations of the virus affect the younger instars more than the older hornworms (Bellotti et al. 1992).

The hornworm baculovirus can be managed by cassava growers. They can collect and macerate diseased larvae and apply the virus suspension to cassava fields. The virus can be stored for several years under refrigeration, and for a few months at room temperature.

Southern Brazil is a "hot spot" for the hornworm. *E. ello* is the most important pest of cassava in the states of Santa Catarina and Paraná. Their agricultural research institutions, Empresa de Pesquisa Agropecuária de Santa Catarina (EMPASC) and Instituto Agronómico do Paraná (IAPAR) have implemented the baculovirus technology for control of cassava hornworm. After solving several technical problems related to handling the virus, EMPASC held field days and training events involving over 400

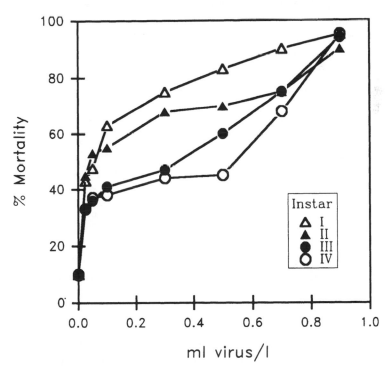

FIGURE 4.1 Effect of baculovirus concentration on mortality of larval instars of *Erynnis ello* (Bellotti et al. in press).

farmers and extension workers. They publicized the use of the hornworm virus on radio and television, airing interviews with farmers who use the technology. They also distributed free baculovirus samples to farmers. Similar activities were carried out in Paraná. In northern Paraná, the principal cassava growing area of the state, 50% of farmers on 34,000 ha were using the baculovirus in 1991 (S. Torrecillas, IAPAR pers. comm.). IAPAR attributed the success primarily to the low cost of the baculovirus (US$1/ha): compared to the cost of chemical control with pyrethroids (US$14/ha). A 60% reduction in pesticide use for hornworm control has been reported in Santa Catarina by EMPASC (R. Pegoraro pers. comm), and this institution has provided the baculovirus to all 25 states in Brazil.

EMPASC wanted to know why the reduction in pesticide use in Santa Catarina wasn't greater, given the low cost of the baculovirus technology. Good management of the hornworm depends on monitoring its presence in the field. Some farmers are better than others at detecting the hornworm at the beginning of an infestation when the hornworms are in the early larval stages. In cases where farmers detected the attack in an advanced stage, they often applied the baculovirus at the recommended dose (0.05 ml virus suspension/l). Under Brazilian conditions, this dose kills the early instars effectively but does not eliminate the fifth instar larvae. Farmers in this situation panicked, and applied pesticides.

This information fed back into the research process. Although less than 50% of fifth instar larvae die at concentrations below 0.45 ml virus suspension/l (Bellotti et al. 1992), most either succumb in the pupal stage or produce deformed adults. Very few adult females emerge and these are sterile. If application of the baculovirus is delayed until the larvae are in the fifth instar, the combination of direct mortality, deformity, and the effect of the baculovirus on the sex ratio interrupts the reproductive cycle of the hornworm and brings the population irruption under control.

EMPASC and IAPAR are mounting educational campaigns to convince farmers that they need not apply pesticides when they see live fifth instar larvae in the field after application of the baculovirus.

The EMPASC/IAPAR program for control of the cassava hornworm in Santa Catarina and Paraná, Brazil rates well when evaluated on the Task Force criteria for successful IPM. EMPASC was heavily involved in the development of the baculovirus technology from its inception, and actively sought the participation of farmers in the testing and adaptation of the technology for local conditions. Santa Catarina farmers have provided valuable feedback which led to important refinements in the technology.

Before the baculovirus became available in southern Brazil, farmers applied pesticides to control cassava hornworm. The competitive price of the baculovirus technology facilitated its adoption by farmers. The technology is within their economic means and technical competence, although the experience with management of late instar larvae shows that farmers revert back to their original practices when they are unsure of themselves. Their technical competence can be improved by well-publicized information updates as the technology evolves, and through refresher training events.

The effect of a technology on labor requirements influences the feasibility of its integration within an existing system, and ultimately, its adoption by farmers (CIAT 1991). The hornworm virus and chemical control require similar investments of labor, facilitating integration of the baculovirus technology with other aspects of cassava crop management.

The Task Force associated success in implementation of IPM with availability of a selection of component technologies. A choice of technology options gives the farmer greater flexibility to incorporate improved technology into current practices. A selection of component technologies is available for management of cassava hornworm. In addition to the inundative release of *Trichogramma* spp. already mentioned, other parasites such as *Telenomus* spp., the larval predators *Polistes* spp., and the bacterial pesticide *Bacillus thuringiensis* can be used successfully, however these require a higher level of decision-making ability on the part of the farmer, or more investment in infrastructure for their production. Recommending these to diversify farmer options may lead to negative experiences, and loss of credibility for biological control. Since all the elements are in place so that IPM of the hornworm in southern Brazil can evolve with experience, diversification may occur in the future when farmers are ready.

Cassava Root Rots

Although pest and disease constraints limit cassava production (Cock 1985), farmers often have little incentive to adopt technology which increases productivity of cassava, unless 1) the new technology is cheaper to implement than the current practices, as in the hornworm case, 2) production problems cause yield to drop below the relatively low, but stable level expected by farmers when the crop is grown for local consumption, or 3) there is a market to absorb the increase in production. An example of the second situation occurred in the early 1980s in the Manaus area of the Brazilian Amazon. Farmers plant cassava during the dry season in the Varzea, the riverbank area that floods during the rainy season. Stable yields of about 20 t/ha were obtained by planting in the highly fertile alluvial soil deposits. In 1983 an epidemic of root rots broke out and yields dropped to 6 t/ha on more than 80,000 ha (Lozano in press).

A network of research and extension organizations (Centro de Pesquisa Agroforestal da Amazonas [CPAA] Centro Nacional de Pesquisa da Mandioca y Fruticultura [CNPMF], Centro Internacional de Agricultura Tropical [CIAT]), and farmers formed to confront the problem. The discovery of sources of host plant resistance in regional germplasm followed the identification of the pathogen complex, (*Phytophthora drechsleri* and *Fusarium solani*). Cultivars developed by crossing resistance sources with local varieties are exchanged within the target area. A set of cultural practice recommendations based on rotation with maize or rice, soil drainage and planting on ridges, selection of vegetative propagation material (stakes), and stake treatment with fungicide were developed and tested with farmer participation. Provided that a tolerant clone is used, the increase in yield from improved technology depends on the number of technology components implemented (as shown in Figure 4.2). On farms where all the technology components were adopted, yields were superior to the pre-epidemic baseline.

As a result of field days and training events, a high proportion of cassava farmers in 40 municipalities use the improved technology. A follow-up study in 3 sites, documented a 90% decrease in root rot incidence, and restoration of yields to the pre-epidemic level.

The Varzea root rot control program is a good example of the benefits of providing a series of component technologies. A formal socioeconomic

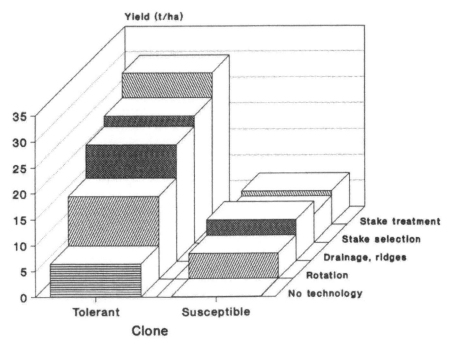

FIGURE 4.2 Cumulative effect of improved technology components for control of fungal root rots on the fresh yield of one tolerant and the mean of 15 susceptible cassava clones in the Amazon Varzea of Brazil.

study of technology adoption will be conducted in 1992, however, preliminary surveys indicate that farmers implement one or more of the technology components depending on how well they fit in with already existing practices. This spontaneous adoption suggests that the technology is economically viable. Some of the root rot control components such as planting on ridges, and selecting and treating of planting material mean increased labor requirements. Despite this, recovering the original stable yield levels has provided sufficient incentive for adoption in this area of limited external marketing possibilities for cassava.

This case illustrates the difficulty of generating widely applicable IPM technology even within a single crop. Cassava root rot management technology validated with farmers in the seasonally dry North Coast region of Colombia (CIAT 1983, 1984, 1985, 1986, 1990, 1991, Lozano in press) was not implementable in the Amazonian Varzea region because of distinct edaphoclimatic conditions and farm characteristics, and a different pathogen complex. Site specific technology components for the Varzea had to be generated locally through interaction between farmers, state (CPAA), national (EMBRAPA) and international agricultural research institutions (CIAT).

Recommendations for Integrated Management of Pest and Disease Complexes in Cassava

Cassava grown in seasonally dry to semiarid lowland areas of the tropics is particularly vulnerable to pest and disease attack. When relative humidity is low, the stomata of cassava close to reduce water loss by transpiration (El-Sharkawy et al. 1985). During prolonged drought, closure of the stomata is not sufficient to prevent water loss, and leaves drop. Because of this decrease in leaf area, dry season pests generally cause the greatest yield losses in cassava (Bellotti et al. 1987). The resting structures of root rot fungi build up to very high levels in the soil, and cause epidemics when activated by rainfall. Epidemics can be particularly severe in regions with low total annual rainfall, but with intense, frequent rain events during the wet season.

In addition to a long history of multiple pest and disease problems, subhumid ecosystems have relatively low resilience to perturbations in water and nutrient cycles. The implementation of farming practices which preserve water resources and maintain soil fertility is particularly critical to agricultural sustainability in these areas. The availability of improved crop production and protection practices is not the main factor limiting the implementation of sustainable farming methods. Poverty spurs environmental degradation in subhumid and other ecologically fragile agricultural lands (Ashby 1985). The identification and opening of markets which can generate rural income, catalyze development, alleviate poverty and motivate farmers to invest in conserving their natural resource base is a critical step in achieving sustainable agriculture.

Cassava can contribute to meeting this challenge. Tolerance of abiotic stresses, such as low soil fertility and drought, contribute to the widespread distribution of cassava in marginal lands and to its reputation as a famine reserve and risk-aversion crop. But this is a limited view of the crop. Its relatively high productivity in difficult environments can be exploited to propel economic development. The roots can be transformed into products as diverse as animal feed, starches and flour, in addition to their use as a fresh or fermented foodstuff (Wheatley and Chuzel 1992).

The high cost of specialized mechanization for harvesting and the relatively high labor requirements for weed control makes cassava a relatively unattractive option for large scale, highly capitalized agriculture. The high labor requirements, the logistics of handling this perishable crop, and the high cost of transporting the bulky roots limit processing operations to sites close to the production area (Cock 1991). Small-scale farmers or farmer associations currently have, and should retain the comparative advantage in small-scale rural agroindustries based on cassava. These add value, generate income, stimulate development, and improve the standard of living in the rural sector (CIAT 1991).

The proliferation of cassava based rural agroindustries leads to intensification of production, which could result in shorter fallow periods, unnecessary or inappropriate use of pesticides, and in the long term, degradation of the environment. Fortunately, the economic incentive to intensify production generates a demand for improved technology and facilitates its adoption, providing a mechanism to avoid or break the cycle of degradation. In Latin America, several of the major cassava growing areas are in subhumid areas of Colombia, Ecuador and Brazil. Cassava production has intensified in these areas as a result of the opening of new markets based on an integrated production, processing and marketing model (Perez-Crespo 1991). As farmer demand for improved production technology grows, ecologically sound crop management and protection practices must be tested and adopted by trained and highly interactive local networks of farmers, extensionists and researchers. These networks should focus on the implementation of crop protection technologies based on host plant resistance, microbial and arthropod biological control agents, and cultural practices, which can be easily and flexibly integrated with currently existing practices, and which are within the economic means and technical grasp of farmers.

The education of local and regional leaders in issues related to providing a favorable policy environment for sustainable agriculture should be initiated as early as possible in any crop protection project. Critical issues include the effects of subsidies for pesticides, and for crops which compete for the same markets as cassava. Public and special interest group concerns about the safety and effectiveness of biological control technology in particular, must also be confronted (Greathead 1991).

Conclusion

A holistic approach is needed if IPM for cassava is to contribute to sustainable agriculture, particularly in fragile subhumid environments. Spontaneous adoption of improved crop production and protection technology can occur if the technology is considerably cheaper than the current practices, or if the sudden onset of a new phytosanitary problem reduces yield to far below the low but stable level expected by farmers. Identification and development of market opportunities for cassava can stimulate adoption of ecologically sound crop protection technology in areas with a history of multiple pest and disease problems, by creating a demand for this technology. Technology components should be developed, tested and modified with active farmer participation throughout the process to ensure their relevance to farmer needs. The effect of the local or regional policy environment, and the concerns of the public and of special interest groups should be considered early in the execution of any crop protection program, since obtaining public endorsement and a favorable policy environment requires an educational campaign.

Acknowledgments

We thank Tom Hargrove for his editorial flair.

Bibliography

Arias B. and A. C. Bellotti. 1984. Perdidas in rendimiento (daño simulado) causadas por *Erinnyis ello* (L.) y niveles criticos de poblacion en diferentes etapas de desarrollo en tres clones de yuca. *Revista Colombiana de Entomologia* 10: 28–35.

Ashby, J. A. 1985. The social ecology of soil erosion in a Colombian farming system. *Rural Sociology* 50: 377–396.

Barbier, E. B. 1989. Cash crops, food crops and sustainability: the case of Indonesia. *World Development* 17: 879–895.

Bellotti A. C. and J. A. Reyes. 1989. "Manejo integrado de plagas insectiles en la agricultura," in *Escuela Agric. Panamericana*. K. L. Andrews and J. R. Quezada (eds.). El Zamorano, Honduras. pp. 490–505.

————. A. R. Braun, J. S. Yaninek, H. R. Herren and P. Neuenschwander. 1987. "Cassava agroecosystems and the evolution of pest complexes." Proc. 11th Int. Congress Plant Protection. Oct. 5–9, 1987. Manila, Philippines. pp. 81–89.

————. C. Cardona and S. L. Lapointe. 1990. Trends in pesticide use in Colombia and Brazil. *Journal of Agricultural Entomology* 7: 191–201.

————. B. Arias and O. L. Guzman. 1992. Biological control of the cassava hornworm *Erynnis ello* (L.) (Lepidoptera: Sphingidae). *Florida Entomologist* (in press).

Bentley, J. W. 1990. Conocimiento y experimentos espontaneos de campesinos Hondureños sobre el maiz muerto. *Manejo Integrado de Plagas* 17: 16–26.

Bentley, J. W. and K. L. Andrews. 1991. Peasants, pests and publications: anthropological and entomological views of an integrated pest management program for small-scale Honduran farmers. *Human Organization* 50: 113–122.

Chambers, R., A. Pacey and L. A. Thrupp (eds). 1989. *Farmer First.* Intermediate Technology Publications, London. 218 p.

CIAT. 1983. *Cassava Program Annual Report 1982-1983.* Centro Internacional De Agricultura Tropical. Cali, Colombia. 521 p.

————. 1984. *Cassava Program Annual Report.* Centro Internacional De Agricultura Tropical. Cali, Colombia. 270 p.

————. 1985. *Cassava Program Annual Report.* Centro Internacional De Agricultura Tropical. Cali, Colombia. 371 p.

————. 1986. *Cassava Program Annual Report.* Centro Internacional De Agricultura Tropical. Cali, Colombia. 254 p.

————. 1990. *Cassava Program Annual Report.* Centro Internacional De Agricultura Tropical. Cali, Colombia. 385 p.

————. 1991. *Cassava Program Annual Report, 1987–1991.* Centro Internacional De Agricultura Tropical. Cali, Colombia. 477 p.

Cock, J. H. 1985. *Cassava: New Potential for a Neglected Crop.* Westview Press. 191 p.

Cock, J. H. 1991. "The development context," in *Integrated Cassava Projects.* Perez-Crespo, C.A. (ed.). Working Document No. 78. Centro Internacional De Agricultura Tropical. Cali, Colombia. pp. 4–16.

El-Sharkawy, M. A., J. H. Cock and A. D. P Hernandez. 1985. Stomatal response to air humidity and its relation to stomatal density in a wide range of warm climate species. *Photosynthesis Research* 7: 137–149.

Greathead, D. J. 1991. Biological control in the tropics: present opportunities and future prospects. *Insect Sci. Appl.* 12: 3–8.

Hatcher, J. E., M. E. Wetzstein and G. K. Douce. 1984. *An economic evaluation of integrated pest management for cotton, peanuts and soybeans in Georgia.* Research Bulletin College Agricultural Experiment Station University of Georgia. No. 318. 28p.

Janzen, D. H. 1985. A host plant is more than its chemistry. *Bulletin Illinois Natural History Survey* 33: 3.

————. D. H. 1986. Biogeography of an unexceptional place: what determines the saturniid and sphingid moth fauna of Santa Rosa National Park, Costa Rica, and what does it mean to conservation biology? *Brenesia* 25/26: 51–87.

————. D. H. 1987. A basis for a general system of insect migration and dispersal by flight. *Nature* 186: 348–350.

Kenmore, P. E. 1988. "Conservation of natural enemies precept, payoff and policy in IPM for tropical rice." Proc. XVIII Int. Congr. Entomol. Vancouver, 3–9

July 1988. Lozano, J. C. Overview of integrated control of cassava diseases. *Fitopatol. Bras.* (in press).

Luckmann, W. H. and R. L. Metcalf. 1975. "The pest management concept," in *Introduction to Insect Pest Management*. R. L. Metcalf and W. H. Luckmann (eds.). John Wiley and Sons, N.Y. pp 3–35.

Napit, K. B., G. W. Norton, R. F Kazmierczak Jr. and E. G. Rajotte. 1988. Economic impacts of extension integrated pests management programs in several states. *Journal of Economic Entomology* 81: 251–256.

NRI. 1991. "A synopsis of integrated pest management in developing countries in the tropics." Synthesis report commissioned by the Integrated Pest Management Working Group. National Resources Institute, Chatham UK. 20 p.

Perez-Crespo, C. A. (ed.). 1991. "Integrated cassava projects." Working Document No. 78. Centro Internacional De Agricultura Tropical. Cali, Colombia. 242 p.

Prokopy, R. J., S. A. Johnson, and M. T. O'Brien. 1990. Second-stage integrated management of apple arthropod pests. *Entomologia Experimentalis et Applicata* 54: 9–19.

Wallner, W. E. 1987. Factors affecting insect population dynamics: differences between outbreak and new outbreak species. *Annual Review of Entomology* 32: 317–340.

Wheatley, C. C. and G. Chuzel. 1992. "Cassava (*Manihot esculenta*) use as a raw material", in *Encyclopedia of Food Science, Food Technology and Nutrition*. Academic Press Ltd., London (in press).

Winder, J. A. 1976. Ecology and control of *Erinnys ello* and *E. alope*, important insect pests in the New World. *PANS* 22: 449-266.

————. J.A. and J. M. de Abreu. 1976. Preliminary observations on the flight behavior of the sphingid moths *Eryinnis ello* L. and *E. alope* Drury (Lepidoptera), based on light trapping. *Ciencia e Cultura* 28: 444–448.

Wolda, H. 1979. *Fluctuaciones estacionales de insectos en el tropico: Sphingidae, Memorias del Congreso de la Sociedad Colombiana de Entomologia.* Julio 25–27, 1979. Cali, Colombia.

5

Effects of Cassava Intercropping and Varietal Mixtures on Herbivore Load, Plant Growth, and Yields: Applications for Small Farmers in Latin America

Clifford S. Gold

Introduction

Cassava (*Manihot esculenta* Crantz) is a tropical root crop and the primary staple for an estimated 750 million people (Cock 1982). In Latin America, a long growing season and low market value have relegated cassava to marginal lands where it is grown by small farmers with limited capital (Sanders and Lynam 1981, Cock 1985). Traditional cassava production has been characterized by intercropping and cultivar mixtures (Lozano et al. 1980).

The agronomy of cassava intercropping and associated land equivalent ratios (used in comparing productivity of multiple and sole cropped systems; Mead and Willey 1980) for cassava systems have been treated by Weber et al. (1979), Leihner (1983), and Mason (1983). Intercropping most often provides more efficient use of resources, greater return on available land, protection against soil erosion and effective weed management (Norman 1974, Mead and Willey 1980). Moreover, short duration intercrops provide an early return in cassava-based cropping systems. However, cassava is a poor competitor in its early stages; although farmers are advised to use additive series intercrops, it is recommended that plant competition be reduced through departures from normal planting arrangements (Leihner 1983) and through regular weeding (Doll 1978).

Diversified agroecosystems, including intercrops, often support lower herbivore load than corresponding monocultures (Altieri and Letourneau

1982, Risch et al. 1983). These reductions have been attributed to increased efficacy of natural enemies (Root 1973, Sheehan 1986) or differences in "resource concentration" among cropping systems (Tahvanainen and Root 1972). However, not all crop combinations bring about reduced herbivore loads and, perhaps more importantly, a given herbivore may show variable responses to the same crop combination over space or time (Andow 1983, Risch et al. 1983).

Most studies on diversified cropping systems and insects have concerned the use of more than one crop species. However, diversification may also be attained through varietal mixtures as well as by intercropping. Genetic diversity in crops is believed to help maintain herbivores at low levels although only limited work has been done in this area (Cantelo and Sanford 1984, Gould 1986a,b, Altieri and Schmidt 1987, Power 1988). Under the small farm conditions in which cassava is most often grown, pest management options are limited. Manipulation of cropping systems provides an important pest management tool which can be readily adopted by small farmers, many of whom already use mixed cropping systems as a means of intensifying production on limited quantities of land. At the same time, breeders, agronomists and extension workers should be aware of potential shifts in pest levels which might follow simplification of existing systems, e.g. following introduction of new varieties which perform better in monoculture.

Pest management represents one element of a cropping system and farmers are more interested in yields of the component crops (and their relative market values) than in insect pest levels per se. Moreover, farmers are most likely to respond to the plant protection characteristics of a cropping system under conditions where pest pressure is a major constraint. Such is the case for cassava in parts of Colombia (whiteflies, burrowing bugs) and across Africa (mealybugs, green mites). Nevertheless, virtually no information has been available on how cassava pests respond to different crop combinations (Gold et al. 1989a).

To investigate the potential for reducing herbivore load and associated yield loss, cassava intercropping trials were conducted at the Departments of Tolima and Cauca, Colombia. Both sites support chronic outbreaks of cassava whiteflies. The objectives of this study were to:

1. Determine the abundance patterns of cassava whiteflies in simple and diversified systems.
2. Assess residual effects (if any) of short duration intercrops on cassava whitefly levels.
3. Evaluate whitefly population dynamics in light of the natural enemy and resource concentration hypotheses.
4. Ascertain the relative effects of intercrop competition and differential herbivore load on cassava growth and yields.

Intercropping systems are complex with numerous interactions between component parts (Parkhurst and Francis 1986). Farming systems are defined by site specific factors (e.g. climate, soils, ecology), farm history (previous crops, soil management), crop parameters (crop combinations, cultivar selection, quality of planting material, relative planting times, crop density, spacing and planting arrangement, pests and diseases) and farmer inputs (soil amendments, weed control). In turn, cropping system texture and host plant quality are likely to influence herbivore colonization, tenure time, survivorship and reproduction.

The specific conditions (weather, planting material quality, inputs) of a given crop cycle may alter the competitive balance between component crops. For example, addition of fertilizers to cassava/cowpea intercrops in phosphorous deficient soils favored cowpea and negatively affected cassava growth rates and yields (Mason 1983). In contrast, yields increased for monoculture cassava with addition of fertilizers at the same site.

As a result, no intercropping trial can be definitive. The complexity of multiple cropping systems and the variability among sites and between crop cycles emphasize the importance of understanding the underlying ecological dynamics which affect herbivore abundance patterns. Such an understanding will be critical if results are to be extrapolated from experimental trials to a broader range of systems and sites.

Methods

Sites

Tolima   Trials were placed on the field station of the Instituto Agropecuario Colombiano (ICA) located in Nataima, Department of Tolima at 400 m a.s.l. The site has two rainy seasons, March—May and September—November, and 1375 mm annual precipitation. Average daily temperatures were 26–30° C. Nataima is at 4° N latitude and daylength varied less than 30 minutes during the year.

Nataima is located in the Magdalena Valley near the base of the central cordillera of the Andes. The site is characterized by large scale commercial farms producing cotton, rice and sesame. These crops receive frequent aerial applications of a wide variety of pesticides (Gold 1987). In contrast, cassava is grown on small farms on the valley floor and in the adjacent foothills.

Cassava whitefly problems date back to at least the mid 1970s and are considered chronic. The dominant species was *Aleurotrachelus socialis* Bondar. *Trialeurodes variabilis* (Quaintance) was also abundant. *Bemisia tuberculata* and *Aleurothrixus floccosus* were present but uncommon.

Cauca   A single trial was placed on a small farm, 5 kilometers west of Pescador, Department of Cauca, Colombia at an elevation of 1525 m. Soils on the 0.48 ha field were acidic and phosphorous deficient (Gold 1987).

Pescador is located in the western cordillera of the Andes. The site is characterized by small farms, frequently on moderate to steep slopes. Dominant crops are cassava, beans and maize. The local cassava variety is used primarily as a source for industrial starch. Whitefly outbreaks, first recorded in 1983, have become a persistent problem. *T. variabilis* comprised more than 99% of the whiteflies observed in this region.

Experimental Design

Three overlapping intercropping trials were planted in Tolima. Trial 1 compared cassava monocultures and cassava/maize intercrops at two densities (additive and substitution designs). Trials 2 and 3 consisted of cassava grown in monocultures (single and mixed variety) and intercropped with cowpea and with maize (additive design); the primary difference between the latter trials was the replacement of a regional cassava variety (Trial 2) with CIAT variety CMC 40 (Trial 3). Trial 4, grown in Cauca, utilized cassava monocultures and cassava/bean intercrops under 3 fertilizer regimes. Methods for trial 2 and 4 are presented here. Further details for all trials are presented in Gold (1987).

Tolima A regional (landrace) variety of cassava (MCOL 2257) was planted in monoculture and intercropped with cowpea (CE—31) and with maize (ICA Compuesto Tropical). A fourth treatment, mixed variety monoculture, consisted of alternate rows of regional cassava and variety CMC 40. All plots (216 m^2) consisted of ten rows (1.8 m apart) of 20 cassava plants (0.6 m apart). Intercrops were planted between the cassava rows (additive design). The treatments were replicated 6 times in a complete block design. In order to partition the effects of intercrop competition and differential herbivore load on cassava growth and yields, whiteflies were controlled in two blocks by biweekly applications of monocrotophos.

Cowpea and maize were harvested 17 weeks after planting (WAP). CMC 40 was harvested 35 WAP and regional cassava was harvested 45 WAP. For purposes of analysis, the cassava cycle was broken into four time periods: Establishment (4–6 WAP), Preharvest (8–16 WAP), Postharvest (18–35 WAP), and Mature (39–45 WAP) where harvest refers to that of the intercrop.

Biweekly egg samples on 25 plants per plot (one leaf per plant) were taken between 4 and 45 WAP. All eggs on half of the central lobe (using alternate sides of the central vein) were counted under a microscope. Leaves (number 4 or 5 from the apex) were chosen from randomly selected plants. Egg density per half lobe was extrapolated to egg numbers per plant (over the 2 week sampling interval) using data for whitefly egg distribution within leaves and leaf production per plant (Gold et al. 1990a).

Analyses (by ANOVA) of egg densities were conducted separately for *A. socialis* and *T. variabilis*. To stabilize the variances, the data for *A. socialis* half lobe populations were transformed to natural logarithms (ln (X+1)) while those for *T. variabilis* were transformed to square roots (X+1/2). Other data did not require transformation.

Cauca A regional variety of cassava, "Algodona," was planted in monoculture and intercropped with beans (*Phaseolus vulgaris* L., Brazilian variety "Carioca"). Each plot (115 m²) contained 200 cassava plants. Three cassava monocultures and three cassava/bean intercrops were grown in each block. These received diammonia phosphate (DAP) (18–46–0) at rates of 100, 250 and 500 kg/ha, respectively (see Gold et al. 1990b for detail). Treatments were replicated three times in a complete block design.

An extended dry period, following planting, resulted in failure of the bean intercrop. This drought, compounded with unusually cold weather, retarded cassava growth, limiting response to applied fertilizers during the study period. Although the original objectives of this trial were compromised, considerable variability in vigor existed between cassava plants within plots. This allowed for examination of whitefly response to individual plants rather than plant assemblages (treatments).

Each week, from 8 to 18 WAP, forty cassava plants per plot were randomly selected and ranked on a scale of 1 to 10 for relative plant size: Largest plants (tallest, most highly ramified, greatest leaf area) were ranked 10, smallest plants were ranked 1, and other plants were graded in between. Adult whitefly densities were ranked on a scale from 0 to 10 for three terminal leaves on the same 40 plants: A score of 10 represented maximal whitefly adult densities observed (over 750 adults per leaf); other scores represented percent of maximal density (9=90%, 8=80%, etc.). Kendall's Tau correlations for nonparametric data were determined between host plant size and whitefly numbers within plots and within blocks.

Results

Host Plant Growth

Intercrop competition negatively affected cassava growth rates (plant height, ramification, leaf production) in the absence of whiteflies (protected plots) (see Gold et al. 1989b for detail). During each trial, cropping system differences in cassava growth were evident by 8–10 WAP. Cassava intercropped with maize etiolated and was taller than cassava in other cropping systems. Smallest plants occurred where cassava was intercropped with cowpea.

During Trial 2, monoculture cassava was more vigorous and produced 35% (protected plots) to 65% (nonprotected) more leaves than cassava which had been intercropped with either cowpea or maize (Gold et al. 1989b). Greater leaf area in monocultures was attributable, primarily, to increased ramification of cassava in this system; at the end of the trial, monoculture cassava had twice as many growing points as cassava which had been intercropped. Moreover, leaf production per terminal was 5–20% greater in monoculture. However, leaf size, nitrogen, phosphorous and potassium (measured in Trials 2 and 4) were similar among cropping systems (Gold 1987, Gold et al. 1990b).

CMC 40, a highly ramified variety, produced 5 times as many terminals and 2.7 times as many leaves as the regional cultivar during Trial 2. Relative growth rates varied by trial; in Trial 2, CMC 40 germinated quicker, had faster early development and shaded the regional cassava in varietal mixtures whereas in Trial 3 the opposite was true.

Cropping system effects on cassava growth rates and architecture afforded whiteflies potential differences in host plant quality. Whitefly abundance patterns, in turn, had effects on continued growth and yield formation of the host plant. These dynamics will be discussed in the following sections.

Whitefly Population Levels

In Tolima, whitefly populations persisted at outbreak levels for all but 8 weeks (during Trial 3) of the study period. Numbers commonly exceeded 200 adults and 10000 eggs per leaf. During Trial 2, density averaged 4229 eggs per leaf (81% *A. socialis*) with peak numbers exceeding 900 adults and 42000 eggs. Total whitefly load averaged 1.35 million eggs per plant between 6 and 35 WAP and probably over 2 million eggs for the entire crop cycle. Although egg densities per leaf peaked between 12–20 WAP, ramification resulted in more terminal leaves and steadily increasing numbers of whiteflies throughout the trial.

Extensive defoliation of cassava plants by the hornworm, *Erinnyis ello* (Trial 3) led to severe drops in whitefly numbers although populations recovered within 4 weeks. During Trial 3, whitefly populations inexplicably crashed at 29 WAP and failed to recover during the remainder of the trial.

In Pescador, whitefly outbreaks (99% *T. variabilis*) occurred in experimental plots from throughout the entire study period (5–18 WAP). Whitefly density averaged 2500 eggs per leaf and population peaks (>1400 eggs/ cm^2) exceeded those in Tolima.

A. socialis and *T. variabilis* displayed strong dispersal capabilities. Cassava planted in land recently cleared of coffee and surrounded by woodland (Melgar, Department of Tolima) was attacked at 5 WAP by *A. socialis* even though the nearest cassava field was 8 km away. In Trials 2 and 3, protected blocks were consistently re-invaded by adult whiteflies necessitating applications of chemical pesticides every two weeks. Moreover, isolated potted plants placed 100 meters from cassava fields were often attacked by both species within 72 hours. These results suggest extensive movement of adult whiteflies between cassava plants and plots.

Whiteflies are weak flyers with limited flight capability; their dispersal is predominantly through passive movement in wind currents. Attack of newly planted, isolated cassava fields suggests the emigration of large numbers of whiteflies from established stands.

Whitefly Abundance Patterns

In Tolima, whitefly load was lower in intercropped systems than in cassava monocultures (Trials 1 and 2). The third trial was confounded by two hornworm outbreaks and initial cropping system differences in white-fly density (lower in intercrop) which disappeared as the trial progressed. In Cauca (Trial 4), where the bean intercrop failed due to drought, whitefly load was equal among systems (Gold 1987).

During Trial 2, intercropping cassava with cowpea significantly reduced whitefly load per leaf (by 46%) and per plant (by 70%), relative to monoculture (shown in Tables 5.1 and 5.2). Effects of cowpea association on *A. socialis* egg densities first appeared at 4 WAP while effects on *T. vari-*

TABLE 5.1 Whitefly eggs per half lobe in regional cassava treatments at ICA-Nataima, Dept. of Tolima, Colombia.

	Time Period				
	Estab.	Prehar.	Posthar.	Mature	Total WAP
Treatment	4-6	8-16	18-35	39-45	4-45
A. *Aeurotrachelus socialis*					
Cassava/cowpea	19b[3]	415c	197b	225b	255b
Cassava/maize	38ab	539b	367a	370a	396a
Cassava mono	48a	652ab	483a	360a	474a
Regional-mix[1]	66a	705a	388a	-	473a
F VALUE[2]	4.75·	13.96··	10.01··	16.46··	8.66··
B. *Trialeurodes variabilis*					
Cassava/cowpea	4	114b	66b	29b	68b
Cassava/maize	16	151a	120a	79a	110a
Cassava mono	7	190a	143a	45ab	131a
Regional-mix[1]	9	159a	48b	-	86b
F VALUE[2]	3.31	4.29·	14.71··	6.21·	12.48··

* P <.05; ** P <.01; WAP: Weeks after planting

[1] Regional-Mix is regional cassava in mixture with cassava variety CMC 40.

[2] *A. socialis* populations log transformed; *T.socialis* square root transformed. Repeated measures ANOVA df (3,9).

[3] Treatments with same letter not significantly different by Duncan's new multiple range test.

TABLE 5.2 Whitefly eggs per plant in regional cassava treatments at ICA-Nataima, Dept. of Tolima, Colombia. Mean population for two week intervals.

	Eggs per Plant (x 1000)		
	Prehar.	Posthar.	Total WAP
Treatment	8-16	18-35	8-35
A. *Aleurotrachelus socialis*			
Cassava/cowpea	40.6b[3]	38.9c	38.6c
Cassava/maize	51.1b	69.4b	57.9b
Cassava mono	85.5a	155.4a	125.9a
Regional-mix[1]	78.3a	83.3ab	81.3ab
F VALUE[2]	12.84··	11.20··	10.74··
B. *Trialeurodes variabilis*			
Cassava/cowpea	10.3b	10.4c	11.1c
Cassava/maize	14.0ab	20.2b	18.1ab
Cassava mono	24.0a	41.6a	37.2a
Regional-mix[1]	16.4a	10.1c	13.9bc
F VALUE[2]	4.37·	15.70··	7.45··

* P <.05; ** P <.01; WAP: Weeks after planting
[1] Regional-Mix is regional cassava in mixture with cassava variety CMC 40.
[2] Repeated measures ANOVA df (3,9).
[3] Treatments with same letter not significantly different by Duncan's new multiple range test.

abilis were first observed at 6 WAP (Gold 1987). Cowpea intercrops had a residual effect on *A. socialis* and *T. variabilis* numbers. At the end of the trial, 28 weeks after intercrop harvest, whitefly egg density (per leaf) for both species was 52% lower in the cassava/cowpea system than in monoculture.

Whitefly load per plant was significantly less (53% lower) in the cassava/maize systems than in monoculture, with greatest reductions following intercrop harvest (Table 5.2). Although whitefly egg density per leaf was 17% lower in intercrops with maize than in monoculture, this difference was not statistically significant (Table 5.1).

In varietal mixtures, regional cassava supported more *A. socialis* eggs per leaf but fewer eggs per plant than did CMC 40 (shown in Table 5.3). In contrast, regional cassava supported higher levels of *T. variabilis* eggs per leaf and per plant than did CMC 40.

Total *A. socialis* load was similar in mixed and regional variety monocultures while *T. variabilis* density was 60% lower in the mixed variety system

TABLE 5.3 Whitefly egg numbers per half lobe and per plant (mean for two week intervals) in cassava varietal mixtures at ICA-Nataima, Dept. of Tolima, Colombia.

WAP	Half Lobe			Plant (X 1000)		
	8-16	18-35	8-35	8-16	18-35	8-35
A. *Aleurotrachelus socialis*						
Regional	705a[2]	388a	501a	78.3b	83.3b	81.5b
CMC40	430b	211b	289b	109.2a	123.6a	118.5a
F VALUE[1]	14.77**	14.78**	12.61**	6.96**	10.98**	10.53**
B. *Trialeurodes variabilis*						
Regional	159a	48a	88a	16.4a	10.1b	12.3b
CMC 40	54b	34b	41b	12.4b	19.3a	16.8a
F VALUE[1]	14.04**	4.24**	32.60**	4.16**	15.70**	13.24**

* P <.05; ** P <.01;   WAP: Weeks after planting.
[1]Half lobe populations of *A. socialis* log transformed;
  T. socialis square root transformed. Repeated measures ANOVA df (1,99).
[2]Treatments with same letter not significantly different by Duncan's new multiple range test.

(Gold et al. 1989d). These results suggest that *A. socialis* did not discriminate between regional cassava and CMC 40 whereas *T. variabilis* displayed a clear preference for the regional cultivar. Moreover, presence of CMC 40 not only lowered absolute numbers of *T. variabilis* within varietal mixtures but also reduced egg density on the regional cassava (Tables 5.1, 5.2). This effect intensified as the cassava canopy closed, suggesting that CMC 40 acted, in part, as a repellent.

Three plots, mixed variety in block 1, cassava/cowpea in block 2, and cassava/maize in block 3, suffered drainage problems (after storms) which resulted in retarded plant growth and leaf production. These plots, containing many stunted plants, supported lower egg numbers of *A. socialis* (27% per leaf, 41% per plant) and *T. variabilis* (38% per leaf, 52% per plant) than the same treatments in other blocks (Gold et al. 1990a).

Natural Enemies

Natural enemies of *A. socialis* included the coccinellid predator, *Delphastus pusillus* Leconte, the platygasterid parasitoid *Amitus aleurodinus* Haldeman and the aphelinid parasitoid *Eretmocerus aleyrodiphaga* (Risbec). *T. variabilis* was attacked by *D. pusillus* and an unidentified parasitoid.

Although *A. socialis* and *T. variabilis* invaded all trials within days of cassava germination, *D. pusillus* adults were slow to enter the same fields.

In Trial 2, for example, *D. pusillus* adults first appeared at 12 WAP, even though the predator was abundant in a previously planted cassava stand, 350 m away. In Trials 3 and 4, *D. pusillus* appeared at 8 WAP although the coccinellid was numerous in established cassava stands 20 meters away. The results suggest that dispersal of *D. pusillus*, following location of a plentiful food supply, is limited.

As a result, during Trial 2, *D. pusillus* was absent or uncommon during the intercrop phase and the period of highest whitefly density (10–20 WAP). *D. pusillus* adult populations peaked between 39–45 WAP (shown in Table 5.4) while larval density was highest at 14–16 WAP and 28–35 WAP. From 18 WAP onwards, the coccinellid displayed a functional response (Gold et al., 1989c) and was more abundant in monocultures than in intercrops.

D. pusillus was most common on leaves containing second and third instar *A. socialis* and second to fourth instar *T. variabilis*. In the laboratory, adults consumed an average of 83 third instar whitefly larvae per day

TABLE 5.4 *Delphastus pusillus* adults and larvae per 300 leaves in regional cassava treatments at ICA-Nataima, Dept. of Tolima, Colombia.

| | Time Period | | | | |
| | Establ. | Prehar. | Posthar. | Mature | Total WAP |
Trial Treatment	4-6	8-16	18-35	39-45	4-45
1. Adults					
Cassava/cowpea	0	20	48b[3]	305b	90
Cassava/maize	0	18	86a	437a	133
Cassava mono0	0	15	133a	367ab	137
Regional-mix[1]	0	21	102a	204c	92
F VALUE[2]	1.00	0.46	8.78**	28.01**	1.84
2. Larvae					
Cassava/cowpea	0	28	34b	5	22
Cassava/maize	0	30	59ab	4	36
Cassava mono	0	9	85a	5	42
Regional-mix[1]	0	23	44ab	3	25
F VALUE[2]	1.00	1.14	4.27*	0.86	1.09

* $P < .05$; ** $P < .01$; WAP: Weeks after planting
[1]Regional cassava in mixture with cassava variety CMC 40.
[2]Repeated measures ANOVA degrees of freedom (3,9).
[3]Treatments with same letter not significantly different by Duncan Multiple Range Test.

(Gold et al. 1989c). Consumption by *D. pusillus* larvae was clearly less than that of adults and was probably on the order of 20 whitefly larvae per day. Predator/prey ratios were similar between treatments and so low that predation appeared to have no effect on whitefly numbers during the intercrop phase and limited effect thereafter (Gold et al. 1989c).

Parasitism of *A. socialis*, detected by exit holes in the host puparia, was first observed midway through the intercrop period and persisted through the remainder of the cropping cycle in each trial (personal observation). In Trial 3 (3 sampling dates), parasitoids emerged from 32% of *A. socialis* pupae. Host feeding and/or unsuccessful parasitism may have been partly responsible for an additional 37% mortality (desiccation) which occurred in the *A. socialis* pupal stage (shown in Table 5.5). Parasitism of *T. variabilis* was negligible (<1%).

Whitefly Survivorship

In Tolima, *A. socialis* survivorship from egg to adult was estimated at 7% (monoculture) to 9% (intercropped systems) (shown in Table 5.6)). *T. variabilis* survivorship could be determined only for the first two instars and was estimated at 50% with no difference among cropping systems (Gold et al. 1991).

Egg mortality, due to hatching failure or predation, was less than 1%. Mortality in the crawler stage was estimated at 15–25% for *A. socialis* and 25–50% for *T. variabilis*. Desiccation and dislodging were probably the major causes of mortality.

Natural enemy impact was estimated at 20% (Table 5.6) and probably

TABLE 5.5 Parasitism of *Aleurotrachelus socialis* pupae on six- week-old cassava leaves by Amitus aleurodinus and *Eretmocerus aleurodiphaga*.

	Cassava/ Cowpea	Cassava/ Maize	Cassava/ Mono
Pupae examined (N)	24024	51273	39090
Whiteflies[1]	30.4%	33.7%	28.9%
Parasitoids[1]	29.9	32.5	34.4
Dead (Desiccated)	39.7	33.8	36.7
Minimum mortality Parasitoids + Dead	69.6	66.3	71.1

[1]Whiteflies and parasitoids identified by shape of exit holes.

TABLE 5.6 Partial life table for *Aleurotrachelus socialis* in different cropping systems, Dept. of Tolima, Colombia.

A. Field wise survivorship

Stage	No. entering each stage	Mortality Factors	Stage No. dying	Generation Rate of Mortality	Rate of Mortality
X	lx	DxF	dx	100 qx	100 rx
Egg	1000	Hatch Failure	10	1.0	1.0
Crawler	990	Disappearance	198	20.0	19.8
Instar 1 & 2	792	Predation	64	8.1	6.4
		Unknown	335	42.3	33.5
Instar 3 & 4	393	Predation	46	11.7	4.6
		Unknown	74	18.8	7.4
		Parasitism	87	22.1	8.7
		Pupal failure	101	25.7	10.1
Adult	85				
		Total mortality			91.5

Eggs per female: 161; Observed sex ratio 1:1; RO: 6.8

B. Survivorship by cropping system

	Cassava mono		Cassava/cowpea		Cassava/maize	
Stage	lx	dx	lx	dx	lx	dx
Egg	679		344		514	
Hatch failure		7 (1)		3 (1)		5 (1)
Crawler	672		341		509	
Disappearance		134 (20)		68 (20)		102 (20)
Instar 1 & 2	538		273		407	
Predation		48 (9)		20 (7)		33 (8)
Unknown		260 (48)		99 (36)		172 (42)
Instar 3 & 4	230		154		202	
Predation		35 (15)		14 (9)		23 (11)
Unknown		28 (12)		38 (25)		37 (18)
Parasitism		57 (25)		30 (19)		46 (23)
Pupal failure		62 (27)		41 (27)		48 (24)
Adult	48		31		48	
RO:	5.8		7.3		7.5	

Numbers in parentheses are percent mortality within stage

never accounted for more than 50% of larval and pupal mortality. Both natural enemy induced and unexplained mortality were similar in intercropped and monoculture systems in spite of differences in host plant architecture, microclimates, host plant quality and whitefly density.

Survivorship in both species fluctuated over time but displayed no apparent relationship with egg density, predator numbers or ambient factors (RH, temperature, rain, wind). Moreover, periods of high or low success of immature whiteflies were not always followed by corresponding shifts in subsequent egg counts. The effects of survivorship on adult populations and subsequent oviposition were confounded, however, by whitefly movement patterns. Nevertheless, estimated survivorship helps explain continuing increase of whitefly populations from already high levels.

Host Plant/Whitefly Interactions

Whitefly distribution among cropping systems and within plots appeared to be related to differences in host plant quality, especially leaf production and canopy cover. In Tolima, whitefly density was highest on more vigorous plants (e.g. monoculture) even though adult and immature numbers were diluted over a greater number of leaves. Moreover, adult whiteflies tended to accumulate on larger plants within plots in Tolima (pers. observation) and in Cauca (shown in Table 5.7; Gold et al. 1990b). Finally, cropping system effects on host plant vigor persisted after intercrop harvest and, therefore, might explain the residual effect of intercrops on whitefly numbers.

TABLE 5.7 Kendall tau correlations for *Trialeurodes variabilis* adult populations (scaled 0 to 10) with cassava plant size (scaled 1 to 10) in monocultures and intercrops with beans[1] for nine sampling dates at Pescador, Dept. of Cauca, Colombia.

Tau	Plot[2]		Total Block[2]
> =.50 62	(38.3%)	11	(40.7%)
.40-.49 47	(29.0%)	10	(37.0%)
.30-.39 26	(16.0%)	4	(14.8%)
.20-.29 18	(11.1%)	2	(7.4%)
<.20 9	(5.6%)	0	
Mean tau	0.44		0.46

[1] Treatments: 3 cassava monocultures and 3 cassava/bean intercrops (additive series) at different fertilizer levels replicated in 3 randomized complete blocks.
[2] N = 40 cassava plants per plot, 240 plants per block.

Whitefly density was normally independent of plant height or terminal location within plants. However, during the final weeks of Trial 2, white-flies were negatively correlated with terminal height above the ground (Gold 1987). At that time, wind damage was observed in higher terminals and, as a result, whiteflies were more abundant on sprawling, etiolated plants in cassava/maize systems.

Cassava Growth and Yields

Cassava in protected plots grew taller and had significantly higher rates of ramification and leaf production per terminal than cassava in non-protected plots. Differences in height and leaf production per terminal were first seen at 14 WAP, coincident with peak whitefly populations per leaf (Gold 1987). Greater ramification in protected plots first appeared at 31 WAP. Whitefly attack reduced leaf life from an average of 12.9 weeks in protected plots to 8.7 weeks in nonprotected plots; reductions averaged 16% in cassava/cowpea systems, 42% in cassava/maize and 31% in cassava monoculture. Greatest reductions in leaf life occurred during peak periods of whitefly attack (Gold et al. 1989b).

In protected plots, cassava intercropped with cowpea produced fewer roots and had lower yields than in other systems (shown in Table 5.8). Yields of regional cassava intercropped with maize, in monoculture and in varietal mixtures were similar. However, with only two replicates, treatment differences were not significant. Cassava intercropped with cowpea supported lower levels of whitefly attack and suffered less yield reduction than other cassava systems (Table 5.8). For example, yield loss in cassava/cowpea intercrops was 13% as opposed to 58% in monoculture. More-

TABLE 5.8 Cassava root yields per plant in different cropping systems at ICA- Nataima, Dept. of Tolima, Colombia.

Treatment	Yield			Root No.			Ave. Wt.Root		
	P	N	Loss[3]	P	N	Loss[3]	P	N	Loss[3]
Cassava/cowpea	1.31	0.1a[4]	0.2b	2.8	2.7	0.1b	.46	.42a	.04b
Cassava/maize	1.90	0.7b	1.2a	4.1	1.8	2.3a	.45	.38b	.08a
Cassava mono	1.90	0.8b	1.1a	4.0	2.3	1.7a	.47	.34b	.13a
Regional-mix[1]	2.10	0.7b	1.3a	4.4	2.1	2.3a	.48	.35b	.13a
F VALUE[2]	2.4	4.4*	29.4**	5.5	2.3	16.8**	.1*	5.1*	8.7**

P: Protected Plots; N: Nonprotected Plots
[1]Regional-Mix is regional cassava in mixture with cassava variety CMC 40.
[2]ANOVA df (3,9).
[3]Yield losses: mean yield difference between non-protected and protected plots.
[4]Treatments with same letter in same column not significantly different by Duncan's new multiple range Test; ANOVA df (3,9).

over, commercial root number was reduced by only 3% in cassava/cowpea intercrops compared to 43% in monoculture. Finally, reduction in root size was 9% in cassava/cowpea association against 28% in monoculture. Yield losses were similar in cassava/maize, cassava monoculture, and mixed cultivar plots.

As a result, cassava intercropped with cowpea out-yielded other cassava systems in the nonprotected environment. Yield differences between treatments, primarily, reflected differences in root size (Table 5.8). Moreover, land equivalent ratios exceeded 1.5 for both intercropped systems in protected and non-protected systems (shown in Table 5.9). Greatest yield advantages were provided by non-protected cassava/cowpea intercrops and by protected cassava/maize systems.

Discussion

Short Duration Intercrops and Cassava Herbivores

Cassava is a perennial shrub, grown as an annual and often interplanted with shorter duration crops. Intercrop harvest creates significant changes in the cropping system environment: total plant density is reduced, plant/ground contrast is increased, exposure to sun and wind are altered, and the arthropod community is likely to be simplified. The direct effects of intercrops on herbivores (eg. repellence, camouflage of host plants, interference in movement, modified microclimates) and their natural enemies (refuges, alternate food sources) will no longer be present.

If the direct effects of intercrops are critical factors in the reduction of cassava herbivores, residual effects would be limited and intercrop influence on population levels would disappear over time after intercrop

TABLE 5.9 Land equivalent ratios of intercropped systems for trial at ICA-Nataima, Department of Tolima, Colombia, April 1984 to February 1985.

Treatment	LER N	P	ATER N	P
Cassava/cowpea	2.15	1.54	1.79	1.12
Cassava/maize	1.57	1.72	1.24	1.35

N: Nonprotected plots
P: Protected plots (biweekly sprays of monocrotophos)
LER: Land Equivalent Ratio
ATER: Area Time Equivalent Ratio assuming two cycles of intercrop per cassava cycle

harvest. Alternatively, intercrop effects on cassava growth and quality are likely to continue for some time into the post-harvest period.

Although reductions in cassava herbivore load during the intercrop period are important, far greater benefits would accrue should such reductions persist into the postharvest period. Cassava has no critical period for yield formation (Bellotti and Schoonhoven 1978, Cock 1978) and yield losses have been correlated with duration of herbivore infestations (CIAT 1982). Postharvest attack is therefore as important as that occurring during the intercrop period.

Cassava Whitefly Population Dynamics

Cropping system effects on insect movement patterns and tenure time are best elucidated by population estimates per unit land (or per plant in stands of equal density). Cassava intercropped with cowpea supported vastly reduced numbers of both A. *socialis* and T. *variabilis* eggs than monoculture cassava throughout the intercrop period. Population differences persisted until the end of the trial, revealing a residual effect of cowpea on whitefly levels. Intercropping cassava with maize also suppressed whitefly numbers although differences were only significant following intercrop harvest.

Increased plant competition in intercropped systems depressed cassava leaf production, thereby, concentrating whitefly populations on fewer leaves. Nevertheless, whitefly density per half lobe was lower in cassava/cowpea systems than in monoculture. Greatest reductions occurred after intercrop harvest, again, demonstrating a residual effect of cowpea on whitefly density. In contrast, intercropping with maize did not affect whitefly density per half lobe.

The natural enemies hypothesis is often advanced to explain reductions in herbivore load in diversified systems. However, lower whitefly populations in intercrops could not be attributed to increased natural enemy impact within these systems. The predator, D. *pusillus*, was of minor importance during the intercrop period and later demonstrated a functional response, being more abundant in monocultures. Parasitism of immature A. *socialis* was equal among cropping systems while parasitism of T. *variabilis* was negligible.

Mortality of immature A. *socialis* attributable to natural enemies (estimated at 20%), disappearance (61%) and pupal failure (10%) was similar among cropping systems in spite of cropping system differences in intercrop architecture, host plant growth rates, canopy cover, and microclimates (wind exposure, relative humidity, shading, temperature). With survivorship equal among cropping systems, intercropping effects on whitefly density which appeared in whitefly egg stages persisted throughout immature development.

The natural enemies hypothesis can, therefore, be rejected in explaining lower populations of whiteflies on intercropped cassava. Moreover, the residual effect of the cowpea intercrop on whitefly populations can not be explained by buildup of natural enemies in this system during the intercrop period.

Alternatively, the resource concentration hypothesis suggests that lower herbivore load in diversified systems may be attributed to differences in host plant location, immigration and emigration rates and tenure time (Pimentel 1961, Tahvanainen and Root 1972, Bach 1980, Risch 1981, Risch et al. 1983, Baliddawa, 1985). Intercrops affect insect movement patterns through interference (physical barriers and confusion of chemical cues), repellency and/or effects on host plant quality.

The data suggest that intercropping affected whitefly population levels, primarily, through changes in host quality, including plant size, leaf area, and, possibly, nutrient content of leaves. These changes, brought about by increased plant competition, persisted through the entire cassava cycle. This would explain the residual effects of short duration intercrops on whitefly population dynamics. Intercrops may also have increased emigration rates from cassava plots (Gold et al. 1990a) but continuous re-invasion of sprayed blocks by large numbers of adult whiteflies, indicating high levels of dispersion, suggests that such effects would disappear after intercrop harvest.

Cassava is a poor competitor in its early growth cycle (Doll 1978, Moody 1986) and many varieties show reduced vigor and yields when intercropped (Leihner, 1983; Okoli and Wilson, 1984). In this study, cassava intercropped with cowpea was smaller and had reduced rates of ramification and leaf production than cassava in other treatments (Gold et al. 1989b). Lower growth rates in cassava/cowpea systems were reflected by lowest yields in sprayed plots where whiteflies were excluded. Cassava intercropped with maize etiolated and displayed less ramification than monoculture cassava; however, these changes did not affect yields.

Whitefly density per half lobe was positively correlated with plots supporting highest rates of leaf production in spite of being diluted over greater leaf area (Gold 1987). For example, relatively low whitefly densities occurred in plots with poor drainage (and stunted plants), suggesting a response to host plant size and/or related differences in nutritional quality. Moreover, greatest numbers of adult whiteflies tended to aggregate on larger plants within plots.

Differences in cassava vigor (and architecture) clearly altered within plant microclimates at oviposition sites in terminal leaves through greater leaf production rates, reduced internodal spaces, and closer terminal proximity to other terminals (shelter effect). Microclimatic differences within the host plant are likely to have greater effects on herbivore tenure time, reproduction and survivorship than intercrop effects on microclimate patterns within the cropping systems.

Differences in cultivar morphology and associated microclimates have been demonstrated to influence *Bemisia tabaci* Gennadius levels in cotton (Ber!inger 1986, Ozgur and Sekeroglu 1986). Cultivars with smaller leaves provided a more open canopy with better air circulation, lower relative humidity and higher air temperatures that was less favorable to *B. tabaci* than cultivars with larger leaves, a more closed canopy and shady, humid conditions favored by this insect. A similar response appears to exist for *A. socialis* and *T. variabilis*.

Cassava often adjusts its growth rate to nutrient availability maintaining a stable nutrient content in its leaves (Cock 1983). Therefore, competition for resources in intercropped systems would likely affect cassava size but not nutrient profiles (Cock pers. comm., Howeler pers. comm.). In both the Tolima and Cauca trials, leaf NPK was equal among treatments (Gold 1987, 1990b). Additionally, in Cauca, leaf nitrogen showed no relationship to levels of whitefly attack. Although cropping systems may affect cassava leaf biochemistry (including secondary plant substances, amino acid profiles, metabolites, and minerals), it appears that physical factors related to host plant size and passive movement in whiteflies may be more critical in determining population levels.

Whitefly movement of greater than a few meters is predominantly passive (Price 1976, Taylor 1984). Cassava plants with greater leaf area (e.g. in monocultures) were probably more efficient at "filtering" out aerial borne whiteflies. These whiteflies would then congregate in leaf terminals presenting higher populations than in plants with lower rates of leaf production.

Moreover, wind speeds decrease within crops creating shelter patches to the lee of individual plants (Pedgley 1982) with lowest wind velocity most likely in plots, eg. monocultures, with larger plants. Such reductions in air currents might cause a greater "fallout" of passively drifting whiteflies (Pedgley 1982) and also reduce the rate of dislodgement of adults from terminal leaves.

Interactions between component crops make intercropping systems inherently complex (Parkhurst and Francis 1986). Intercrop conditions favoring or discouraging insect buildups may not be intrinsic to a crop combination but depend on the competitive balance between component crops and its effect on host plant quality in a given crop cycle.

If host plant quality is important, the interaction between crop species as influenced by sites and conditions specific to the crop cycle, rather than the crop combination per se, may be the most critical factor in determining herbivore population sizes. This could explain the variable response of some insects to cropping systems.

Cassava Growth and Yield: Intercropping Implications

Cassava is frequently grown in mixtures with short duration intercrops (e.g. maize, beans, cowpea, groundnut) which are present for 30–40% of the cassava cycle. Land equivalent ratios commonly show cassava based intercropping systems to be highly productive, suggesting a more efficient use of resources especially early in the cycle (Leihner 1983). However, little information is available to partition the effects of intercrop competition and differential herbivore load on cassava yields.

Maximal cassava growth and yields were expected in protected monocultures where plant competition and herbivore pressure were at low levels. Growth parameters were, in fact, greatest in monoculture plots although yields were similar in monoculture and cassava/maize intercrops. In contrast, yield differences between protected and nonprotected plots were far more pronounced than the differences in ramification and leaf production. This suggests that whiteflies acted as a nutrient sink without affecting above ground growth.

By comparison, intercrop competition and shading by the quicker growing CMC 40 did reduce above ground vigor (including ramification and leaf production) of regional cassava in both protected and non-protected environments. Cowpea, climbing directly on cassava plants, led to stunted plants. In contrast, shading by maize intercrops resulted in etiolated cassava.

Cassava yield is a function of root number and root size. Although cassava is reported to have no critical period for yield formation (Bellotti and Schoonhoven 1978, Cock 1984), root number may be affected by stress occurring early in the cassava cycle (Hunt et al. 1977). Reductions in root number are usually compensated for by increased root size, without loss of yield, provided root number does not fall below 9 per plant (Hunt et al. 1977, Cock 1978).

In this study, limited replicates precluded meaningful statistical comparisons of yields in systems where herbivores were excluded. Nevertheless, the data suggest that early in the cassava cycle competition was more intense with cowpea than with maize; as a result, intercropping with cowpea reduced yields, through a decrease in root number, while intercropping with maize did not.

Additionally, leaf life is also a critical factor in root production; reductions in leaf life can cause serious yield loss (Cock 1978, 1984, Cock et al. 1979). In the absence of whiteflies (protected plots), intercropping with cowpea reduced leaf life and this, undoubtedly, contributed to lower yields. In contrast, leaf life was greatest in cassava/maize systems.

Density of herbivores per leaf is probably the best indicator for potential yield loss for crops in general (Southwood 1978) and cassava in particular (Cock pers. comm.). In this study, yield loss closely paralleled trends in whitefly density per half lobe with lowest levels of whiteflies and yield reductions in cassava/cowpea systems. Yield losses were due, primarily, to decreases in root numbers and, secondarily, to reductions in root size. Cassava intercropped with cowpea demonstrated the least damage for all growth and yield parameters measured.

Higher population levels of cassava whiteflies were associated both with plots containing more vigorous plants and with more vigorous plants within plots (Gold 1987). The question arises as to why greater whitefly numbers in monocultures did not act as a negative feedback mechanism, reducing vigor, or, alternatively, why yields for these more vigorous plants were not greater.

Cassava partitions its assimilates among root storage, stem development and leaf production. Under stress conditions, above ground parts are often favored over storage roots (Hunt et al. 1977, Cock et al. 1979, Cock 1983). Thus, monoculture cassava maintained vigorous above ground growth at the expense of root production under conditions of intensive whitefly attack. At the same time maintenance of above ground vigor probably continued to encourage greater herbivore numbers in this system.

In summary, cowpea intercrops provided high levels of competition, reducing growth rates in cassava. This, in turn, made the cassava less attractive (or suitable) to whiteflies and greatly reduced yield losses, relative to other treatments. As a result, cassava associated with cowpea out-yielded cassava in the other systems. These data demonstrate that under conditions of high insect attack, reductions in herbivore numbers can more than offset the negative effects on yield brought on by intensive interspecific competition between intercrops.

Application for Small Farmers

In Latin America, cassava is grown by resource poor farmers with limited pest control options. The crop is attacked by a complex of insects and mites; some are seasonal (e.g. green mites) while others are present throughout the year (e.g. whiteflies).

Farmer options for responding to pest problems are limited. Viable alternatives for cassava pest management include cultural controls (including cropping system management), use of natural enemies, and introduction of resistant varieties. It is unlikely that cassava market prices will ever justify chemical controls or other costly inputs.

Cassava is a perennial shrub treated as an annual; its crop cycle is normally 10 to 16 months but may extend to 24 months (Cock, 1985). The long growing season insures that cassava is present throughout the year. Depending on local conditions and farming methods, its age structure may be uniform or highly variable.

Cassava is not known in the wild; congeneric species, however, are generally scattered in small populations. Based on its chemical defenses and perennial phenology, cassava does not fit easily into either the apparent or unapparent categories sensu Feeny (1976). Nevertheless, most important pests in the Americas, where cassava is endemic, are specialists. Intercropping has been shown to be most often effective in reducing population densities of specialists herbivores (Risch et al. 1983). Moreover, many cassava growers are predisposed to intercropping, suggesting that they might be amenable to modifications of existing cropping systems.

The role of natural enemies in controlling cassava pests in Latin America has not been fully explored. In some cases, it has been hypothesized that cassava herbivores may have attained pest status following range extensions (*T. variabilis* in Cauca) or disruption of existing natural controls due to insecticide drift (*A. socialis* in Tolima) or changes in farming practices (Gold 1987). However, the potential use of natural enemies in these situations is unclear and may be limited.

Cassava breeders have successfully developed varieties resistant to thrips and have produced cultivars demonstrating resistance to whiteflies and other pests. However, to date, improved varieties have not been widely accepted in Latin America. Cassava is grown across a wide range of ecological zones and landrace varieties may be better adapted to local conditions. Nevertheless, data on *T. variabilis* response to varietal mixtures suggest that incorporation of resistant cultivars into systems containing agronomically superior landrace varieties may lead to reductions of herbivore levels.

In this context, cultural controls may be the most efficient means of controlling cassava pests. Manipulation of cropping systems provides an ecologically based front line of defense which can serve to discourage herbivore buildups by reducing colonization, increasing emigration, or augmenting mortality rates through enhanced natural enemy action or changes in host plant quality (Altieri and Letourneau 1982, Risch et al. 1983, Andow 1991). Nevertheless, the current study shows that not all intercrop combinations are equally effective at reducing cassava herbivore levels and increasing yields. Variability in insect response to crop combinations has also been demonstrated for cassava hornworms and stemborers (Gold et al. 1990c). These results emphasize the need to understand the ecological mechanisms underlying pest response to cropping systems. This information is not available for most cassava herbivores.

The present study illustrates the dynamics of cropping system—herbivore interactions and the potential use of mixed cropping in controlling pests and increasing productivity (through reduced yield loss and high land equivalent ratios). Diversified systems were demonstrated to substantially reduce herbivore levels; moreover, the intercropping effects transcended the intercrop period and persisted throughout the cassava cycle. Such residual effects are critical in the management of dry season cassava pests (e.g. green mites) which build up after intercrop harvest, when cassava remains in effective monoculture. Finally, yield losses were cut by 80% and land equivalent ratios for the intercrops displayed considerable yield advantages over monoculture cassava.

Conclusion

In summary, crop pests have demonstrated a wide range of responses to diversified cropping systems. As a result, it is impossible to make broad generalizations about the effects of intercropping on herbivore load and associated yield losses. However, the tendency is for intercropping systems to reduce levels of specialist herbivores (Risch et al. 1983). Moreover, numerous other agronomic advantages are gained by intercropping (Willey 1979a, b).

In Tolima, lower herbivore load and associated reductions in yield loss more than offset the negative effects of intercrop competition on cassava growth rates and yields (Gold et al. 1989b). Lower cassava whitefly densities in intercrops were attributed to the smaller size of host plants found in these systems. Yields, however, were not correlated with plant size. Therefore, the relationship between plant size and herbivore attack should be considered in both management of cropping systems and breeding strategies for improved varieties.

Bibliography

Altieri, M. A. and D. K. Letourneau. 1982. Vegetation management and biological control in agroecosystems. *Crop Protection* 1: 405–430.

Altieri, M. A. and L. Schmidt. 1987. Mixing broccoli cultivars reduces cabbage aphid numbers. *California Agriculture* 41: 24–26.

Andow, D. 1983. "Plant diversity and insect populations: Interactions among beans, weeds, and insects." Unpubl. Ph. D. dissertation. Cornell University. 201 p.

————. 1991. Vegetational diversity and arthropod population response. *Annual Review of Entomology* 36: 561–586.

Bach, C. E. 1980. Effects of plant density and diversity on the population dynamics of a specialist herbivore, the cucumber beetle, *Acalymma vittata* (Fab.). *Ecology* 61: 1515–1530.

Baliddawa, C. W. 1985. Plant species diversity and crop pest control. *Insect Science and its Application* 6: 479–487.

Bellotti, A. C. and A. van Schoonhoven. 1978. Mite and insect pests of cassava. *Annual Review of Entomology* 23: 39–67.

Berlinger, M. J. 1986. Host plant resistance to *Bemisia tabaci*. *Agriculture, Ecosystems and Environment* 17: 69–82.

Cantelo, W. W. and L. L. Sanford 1984. Insect population response to mixed and uniform plantings of resistant and susceptible plant material. *Environmental Entomology* 13: 1443–1445.

CIAT. 1982. *Annual Report for 1981*. Cali, Colombia.

Cock, J. H. 1978. "A physiological basis of yield loss in cassava due to pests," in *Proceedings Cassava Protection Workshop*. T. Brekelbaum, A. C. Bellotti and J. C. Lozano (eds.). CIAT. Cali, Colombia. pp. 9–16.

————. 1982. Cassava: A basic energy source in the Tropics. *Science* 218: 755–762.

————. 1983. "Cassava", in *Potential Productivity in Field Crops Under Different Environmental Conditions*. S. Yoshida (ed.). IRRI. Los Banos, Philippines. pp. 341–349.

————. 1984. "Cassava", in *The Physiology of Tropical Root Crops*. P. R. Goldsworthy and N. M. Fisher (eds.). Wiley and Sons, New York. pp. 524–529.

————. 1985. *Cassava: New Potential for a Neglected Crop*. Westview Press, Boulder, Co.

Cock, J. H., D. Franklin, G. Sandoval and P. Juri. 1979. The ideal cassava plant for maximal yield. *Crop Science* 19: 271–279.

Doll, J. 1978. "Weeds: an economic problem in cassava." in *Proceedings Cassava Protection Workshop*. T. Brekelbaum, A. C. Bellotti and J. C. Lozano (eds.). CIAT. Cali, Colombia. pp. 65–69.

Feeny, P. 1976. Plant apparency and chemical defense. *Recent Advances in Phytochemistry* 10: 1–40.

Gold, C. S. 1987. "Crop diversification and tropical herbivores: effects of intercropping and mixed varieties on the cassava whiteflies, *Aleurotetrachelus socialis* Bondar and *Trialeurodes variabilis* (Quaintance), in Colombia." Unpubl. Ph.D. dissertation. Univ. of Calif., Berkeley. 362 p.

Gold, C. S., M. A. Altieri and A. C. Bellotti. 1989a. Cassava intercropping and pest incidence: A review illustrated by a case study in Colombia. *Tropical Pest Management* 35: 339–344.

————. 1989b. Effects of intercrop competition and differential herbivore numbers on cassava growth and yields. *Agriculture, Ecosystems and Environment* 26: 131–146.

————. 1989c. The effects of intercropping and mixed varieties on predators and parasitoids of cassava whiteflies in Colombia: An examination of the "natural enemies hypothesis." *Bulletin of Entomological Research* 79: 115–121.

————. 1989d. Effects of varietal mixtures on the cassava whiteflies, *Aleurotrachelus socialis* Bondar and *Trialeurodes variabilis* (Quaintance) in Colombia. *Entomologia experimentalis et Applicata* 53: 195–202.

————. 1990a. Direct and residual effects of short duration intercrops on the cassava whiteflies *Aleurotracelus socialis* and *Trialeurodes variabilis* in Colombia. *Agriculture, Ecosystems and Environment* 32: 57–67.

————. 1990b. Response of the cassava whitefly, *Trialeurodes variabilis* (Quaintance) (Homoptera: Aleyrodidae) to host plant size: implications for cropping system management. *Acta Oecologica* 11: 35–41.

————. 1990c. Effects of intercropping and varietal mixtures on the cassava hornworm, *Erinnyis ello* (Lepidoptera: Sphingidae), and the stemborer, *Chilomima clarkei* (Amsel) (Lepidoptera: Pyralidae) in Colombia. *Tropical Pest Management* 36: 362–367.

————. 1991. Survivorship of the cassava whiteflies, *Aleurotrachelus socialis* and *Trialeurodes variabilis* (Homoptera: Aleyrodidae) under different cropping systems in Colombia. *Crop Protection* 10: 305–309.

Gould, F. 1986a. Simulation models for predicting durability of insect-resistant germplasm: A deterministic diploid, two-locus model. *Environmental Entomology* 15: 1–10.

————. 1986b. Simulation models for predicting durability of insect-resistant germplasm: Hessian fly (Diptera: Cecidomyidae)-resistant winter wheat. *Environmental Entomology* 15: 11–23.

Hunt, L. A., D. W. Wholey and J. H. Cock. 1977. Growth physiology of cassava. *Field Crops Abstracts* 30: 77–91.

Leihner, D. E. 1983. *Management and Evaluation of Intercropping Systems with Cassava.* CIAT, Cali, Colombia. 70 p.

Lozano, J. D., D. Byrne and A. Bellotti. 1980. Cassava/ecosystem relationships and their influence on breeding strategy. *Tropical Pest Management* 26: 180–187.

Mason, S. C. 1983. "Land use efficiency, canopy development, dry matter production, and plant nutrition of cassava-grain intercropping." Unpubl. Ph.D. dissertation. Purdue University.

Mead, R. and R. W. Willey 1980. The concept of "land equivalent ratio" and advantages in yields from intercropping. *Experimental Agriculture* 16: 217–228.

Moody, K. 1986. Weed control in cassava, a review. *Journal of Plant Protection in the Tropics* 2: 27–40.

Norman, D. W. 1974. Rationalizing mixed cropping under indigenous conditions: the example of northern Nigeria. *Journal Development Studies* 1: 3–31.

Okoli, P.S.O. and G. F. Wilson. 1984. Response of cassava (*Manihot esculenta* Crantz) to shade under field conditions. *Field Crops Research* 14: 349–359.

Ozgur, A. F. and E. Sekeroglu. 1986. Population development of *Bemisia tabaci* (Homoptera: Aleyrodidae) on various cotton cultivars in Cukorova, Turkey. *Agriculture, Ecosystems and Environment* 17: 83–88.

Parkhurst, A. M., and C. A. Francis. 1986. "Research methods for multiple cropping systems," in *Multiple Cropping Systems.* C. A. Francis (ed.). MacMillan, London. pp. 285–316.

Pedgley, D. 1982. *Windborne Pests and Diseases: Meterology of Airborne Organisms.* Ellis Horwood Ltd., Chichester. 250 p.

Pimentel, D. 1961. Species diversity and insect population outbreaks. *Annals of the Entomological Society of America* 54: 76–86.

Power, A. G. 1988. Leafhopper response to genetically diverse maize stands. *Entomologia experimentalis et Applicata* 49: 213–219.

Price, P. W. 1976. Colonization of crops by arthropods: non-equilibrium communities in soybean fields. *Environmental Entomology* 5: 605–611.

Risch, S. J. 1981. Insect herbivore abundance in tropical monocultures and polycultures: an experimental test of two hypothesis. *Ecology* 62: 1325–1340.

Risch, S. J., D. Andow and M. A. Altieri. 1983. Agroecosystem diversity and pest control: Data, tentative conclusions, and new research directions. *Environmental Entomology* 12: 625-629.

Root, R. B. 1973. Organization of a plant-arthropod association in simple and diverse habitats: the fauna of collards (*Brassica oleracea*). *Ecological Monographs* 43: 95–124.

Sanders, J. H. and J. K. Lynam. 1981. New agricultural technology and small farmers in Latin America. *Food Policy* 6: 11–18.

Sheehan, W. 1986. Response by specialist and generalist natural enemies to agroecosystem diversification: a selective review. *Environmental Entomology* 15: 456–461.

Southwood, T.R.E. 1978. *Ecological Methods*. Chapman and Hall, London. 524 p.

Tahvanainen, J. O. and R. B. Root 1972. The influence of vegetational diversity on the populational ecology of a specialized herbivore, *Phyllotreta cruciferae* (Coleoptera: Chrysomelidae). *Oecologia* 10: 321–346.

Taylor, L. R. 1984. Assessing and interpreting spatial distributions of insect populations. *Annual Review of Entomology* 29: 321–357.

Weber, E., B. Nestel and M. Campbell 1979. Intercropping with cassava. IDRC, Ottawa. 142 p.

Willey, R. W. 1979a. Intercropping: its importance and its research needs. Part I. Competition and yield advantages. *Field Crops Abstracts* 32: 1–10.

Willey, R. W. 1979b. Intercropping: its importance and its research needs. Part II. Agronomic relationships. *Field Crops Abstracts* 32: 73–85.

6

Human Interactions in Classical Biological Control of Cassava and Mango Mealybugs on Subsistence Farms in Tropical Africa

Peter Neuenschwander

Introduction

In a period of political and economic turmoil in Africa, when per capita food production is sinking and production and population trends are generally unfavorable (Michler 1991), maintaining or developing sustainable agriculture (Weil 1990) should and often has become the highest priority for policy makers and researchers alike. Small holders represent a large proportion of Africa's population. Their crop protection strategies such as burning, use of crop diversity, intercropping, use of genetically resistant crop varieties, and weed control practices, have recently drawn attention (Hussey 1990, Kirkby 1990) and it is now understood that any new research results must fit into a traditional agroecosystem in order to be adopted by the farmer.

The guiding principle in crop protection in the last thirty years has been the concept of integrated pest management (IPM) as defined by Stern et al. (1959), but outlined previously (De Bach 1951). World-wide, many successful examples can be listed (Huffaker 1980; Croft et al. 1984). For cassava, the major subsistence staple food in Africa (Okigbo 1989) and a main topic of this text, individual, often untested, pest management techniques have been listed rather indiscriminately (Hahn et al. 1979; Gahukar 1991). Among them, resistance breeding has a long history (Beck 1980; Hahn et al. 1989) and, in view of the important role of moderate degrees of resistance (Wilbert 1980), good results against major diseases have been achieved. Agronomic practices in intercrops, which most often include cassava, provide another tool, but are only rarely aimed at reducing pests

143

other than weeds (Steiner 1982, Akobundu 1990). In most conditions of small holder farms, insecticides are no option against well-established pests because of their relative inefficiency, side-effects, and difficulty in proper timing of the applications. In reality, insecticides are not used because the chemicals and the equipment to spray insect pests are often not available or too expensive for the small holder to use on cassava, a crop of low cash value.

Insecticide use, and reduction of existing insecticide misuse in order to preserve natural enemies, is inherent in IPM. However, severe complications arise with this concept, particularly on non-cash crops. In this case, economic thresholds can often not be established because insecticide treatments are not a viable option. Even where they can be calculated, they tend to increase during the season, thus adding another complication (Hueth and Regev 1974). Plants also are influenced by insecticide applications which can have the effect of foliar fertilization (Chaboussou 1970). More important, insect pests resurge from the destruction of their natural enemies. Where this is the case, better timing of the insecticide applications cannot bring them back. Insecticide use often begets more insecticide use, a spiral that can only be broken by a political decision (Carson 1962, van den Bosch 1978; Gips 1987). Selective use of insecticides is usually given as an answer, and it is indeed feasible (Greathead 1989a, Pickett 1988). However, the record in safety and efficiency in the use of pesticides among smallholders is mostly poor (Andrews and Bentley 1990). Moreover, the often applied state controlled subsidies in developing countries are incompatible with IPM (Goodell 1984). In a recent review on IPM in Africa (Kiss and Meerman 1991), insecticides were only applied on cash crops and the successful programs on food crops all worked without economic thresholds and chemical interventions. It is concluded that IPM, as practiced in Europe, America, and some parts of Asia, has little relevance for smallholder agriculture in Africa.

This unease with insecticide management, combined with often misdirected resistance breeding (van Emden 1991), has gradually led to the development of a more holistic approach (Huffaker 1974, Croft et al. 1984). This culminates in the definition of systems management, which puts emphasis on prevention by repairing agricultural ecosystems rather than relying only on what is still present (Delucchi 1987, 1989). This approach requires a study of the entire ecosystem, from where the agricultural system was derived, including soils (Lal 1982), plant genetic resources (Oldfield and Alcorn 1987), and all associated organisms on different tropic levels. It is based on an understanding of how undisturbed insect populations are regulated (Huffaker et al. 1984, May and Hassell 1988), how the individual insects behave among each other and in relation to their plant hosts (Kennedy 1977, van Alphen and Vet 1986, Heinrichs 1988), and how they fulfill their nutritional and other requirements (Hagen 1986), such as alternate food sources and hosts (Doutt and Nakata 1965, Hagen 1976, McMurtry and Rodriguez 1987, Vinson and Barbosa 1987). Before this knowledge can be put into practice, however, much more needs to be known about the importance of the various factors driving and influencing complicated farming systems, especially in the tropics (Edwards 1989).

Including man into this system puts the farmer into a central place and adds another dimension to systems management (Walker and Norton 1982). In view of farmers' experiments with their crops since time immemorial (Haverkort 1991), the purely top-down approach to improve agriculture, i.e. from research to extension to farmer, must be viewed as a costly anachronism (Weiss and Robb 1989). Development in the direction of increasing farmer cooperation in research culminates in agroecology, which seeks to enhance agricultural productivity to benefit rather than marginalize small producers (Altieri 1987, Hecht 1987, Altieri and Hecht 1990). While the old farming systems research emphasized technology transfer, agroecology seeks to define principles, leaving specific technological forms to be determined by the local milieu. In its extreme, this approach rejects the philosophy of research as developed in the western world (Norgaard 1988) and removes it altogether from problem solving. Whatever weights are given to the different partners responsible for agricultural research output, the fact remains that the envisaged creation of sustainable systems with regional conditions of near equilibrium asks for more research, education, information storage and retrieval, and often a change in the economic, political, and legal framework (Harmsen 1990).

In the present paper, we shall try to place the rather staid concept of classical biological control into the framework of systems management and agroecology. Both aspects, the science and technology development and the socio-economic realities will be discussed with the example of two classical biological control projects on mealybugs (Homoptera, Pseudococcidae) by two endophagous parasitoids (Hym., Encyrtidae) in Africa. One involves the successful biological control of the cassava mealybug (CM), *Phenacoccus manihoti* Mat.-Ferr. by the parasitoid *Epidinocarsis lopezi* (De Santis), which by 1991 was established in 26 African countries. The technological aspects have recently been reviewed (Herren and Neuenschwander 1991). The second example concerns biological control of the mango mealybug (MM), *Rastrococcus invadens* Williams, by the parasitoid *Gyranusoidea tebygi* Noyes, a very close relative of *E. lopezi*, established in eight West and Central African countries.

From these examples, we hope to draw general conclusions for classical biological control, as defined by Garcia et al. (1988), which aims to reestablish the population equilibrium found in the area of origin (Croizat et al. 1974). To achieve this, natural enemies that are thought to be responsible for control are searched during foreign exploration (Zwölfer et al. 1976), studied and tested in quarantine, reared, and introduced where an accidentally introduced organism has become a pest. The discussion thus excludes other sorts of biological control (Greany et al. 1984, Hoy 1985, 1988) and does not concern the biological control of native pests by introduction of natural enemies from related hosts (Pimentel 1963, Hokkanen and Pimentel 1984). Achievements and failures of classical biological control in Africa have been reviewed (Greathead et al. 1971), new opportunities have been pointed out (Greathead and Waage 1983, Greathead 1989b), and a short review of the entire Biological Control Program of the International Institute of Tropical Agriculture (IITA), of which these two mealybug projects are part, is given by Herren (1990). The present review therefore highlights historical aspects, which do not find a place in a strictly entomological review, and sociological problems, which have received only glancing mention in the literature on classical biological control. The human interactions observed in these two projects will be described, providing material for an ultimate study by a sociologist.

History of the Projects: Identification of the Pest Problem, Plan of Action, Collaborators, and Administrative Set-Up

Cassava Mealybug

For a successful solution, the problem first has to be clearly identified, hence the priority for crop loss assessment. In the case of the CM, however, the damage is so devastating and obvious that, after its discovery in 1973 in the Kinshasa/Brazzaville area in central Africa, no serious attempt at crop loss assessment seemed necessary. The general outcry by farmers, scientists, and politicians alike was for an international meeting in M'Vuazi, Zaire. There biological control and resistance breeding was to be undertaken by IITA in collaboration with other institutions, to choose long-term instruments to achieve a sustainable solution (Nwanze and Leuschner 1978). Some resistance against the CM was later found in several IITA varieties (Hahn et al. 1989), but the most important contribution to solving the CM problem was to come from classical biological control. Search for natural enemies was planned in South America, the perceived center of evolution of cassava (Renvoize 1973). From the start, the Inter-African Phytosanitary Council (IAPSC) of the Organization for African Unity, a unique continental umbrella organization, though without legislative power, was to be involved. Its guidelines (O.A.U. 1988) were respected.

Following an appeal by the Zairian government in 1979, IITA hired a biological control specialist whose first success was to coordinate and finance, collaboration between the sister institute in Colombia, CIAT, the Brazilian EMBRAPA, the now IIBC in London, the Nigerian quarantine, and the IAPSC, whose head had to approve the quarantine facilities in

England. A coordinated information campaign by the IAPSC was considered important because both the CM and later *E. lopezi* spread across political boundaries, which in Africa are mostly artificial. Over many years, acceptance by quarantine authorities all over Africa was obtained to import beneficial insects into their respective countries. Each country thereby was autonomous in its decisions, free to accept or reject the sometimes ambiguous recommendations from the IAPSC. Though countries were advised to report releases to the IAPSC, they often did not do so. Recently, however, communications improved thanks to yearly meetings of IAPSC and IITA staff to discuss planned releases in the different countries.

Once the main fear about safety had been alleviated (Caltagirone and Huffaker 1980), the main bone of contention in these discussions with the IASPC and the quarantine authorities of the individual countries was the function of quarantine. In most countries, our view prevailed that quarantine could only guarantee non-noxiousness of the exotic organism and that its efficiency had to be tested in test releases. The appealing, but technically impossible, request to test exotic organisms in various simulated ecological conditions did not prevail, but the idea lingers on, as does the dangerous perception that quarantine should be done in Africa. Even IIBC in its Kenya station does not perform incoming quarantine (M. Cock, pers. comm.), simply because the danger for mishaps in the conducive tropical climate is too big.

The next step was to intensify and coordinate foreign exploration. At the same time, rearing facilities were established in Nigeria, first near Abeokuta, because the CM had not spread to the IITA campus yet, and later at IITA Ibadan. This set the basis for test releases, which were done only after receiving a request by the affected countries and always in close collaboration with the national authorities. But even after a quarantine permit and a request for technical assistance had been issued to IITA by a country, problems of coordination often persisted because the quarantine authorities in most countries are in other branches of the same ministry or even in other ministries than research. Since any initial test releases called for research, coordination among the entomologists of different national organizations had to be assured. For this purpose, the training project, which was started by the biological control program and later expanded and supported by the Food and Agriculture Organization (FAO) with United Nations Development Programme (UNDP) financing, advertised and stimulated the creation of so-called national biological control programs, which would encompass all national scientists concerned (Wodageneh 1989). In Nigeria, from the start, IITA entomologists were members of the National Biological Control Committee, which included quarantine services, the National Root Crops Research Institute, various universities, and other agencies.

As the cassava mealybug and mite biological control project, now named Africa-wide Biological Control Program, was growing, various European donor agencies supported and continue to support the work in all its aspects. Since these biological control activities were new to IITA, the donors established an international Expert Advisory Committee (EAC) to guide the work of the IITA entomologists. The EAC met yearly, and an additional mid-term evaluation was undertaken in 1987 (Arnold et al. 1987).

Establishing collaboration between technical agencies (IITA, CIAT, IIBC, national institutions), regulatory agencies (IAPSC, national quarantine services), and donors (united in a donors' group) must be considered the key to the future success of this classical biological control program. All releases were done in collaboration with the national institutions and were often accompanied by interviews and reports in local newspapers, supported by pamphlets written by IITA. Thus the idea of classical biological control was spread also to the farmers and consumers and the awareness of the value of classical biological control was heightened in Africa.

This led eventually to a reorganization of IITA and to the move of the entire program to Cotonou in the Republic of Benin into the newly constructed Biological Control Center for Africa in 1988. Since 1992, the Biological Control Program has been placed into the newly created Plant Health Management Division with headquarters in Cotonou, one of three Divisions conducting research in IITA. Thus plant protection has finally gained the administrative place it deserves, considering that losses due to pests (insects, nematodes, diseases, and weeds) amount to about 30% of potential global production (NRI, 1991), and biological control in conjunction with ecological studies is recognized as its main pillar.

Mango Mealybug

The MM was discovered as a serious pest on numerous fruit trees and ornamentals in Togo and Benin, and described as a new species, presumably of South-East Asian origin (Williams 1986, Agounké et al. 1988). IIBC immediately started foreign exploration, which led to the discovery of promising parasitoids (Narasimham and Chacko 1988). A biological control project was initiated by FAO at the Plant Protection Services in Cacaveli, Togo and Porto-Novo, Benin, both sponsored by the German "Gesellschaft für Technische Zusammenarbeit" (GTZ). A seminar, similar to the one held in M'Vuazi, was held in Lomé (Séminaire-atelier 1987), where an OAU regional project was created with defined roles for IIBC and IITA. The national biological control programs were already in place and financially supported by the CM biological control project. Classical biological control was considered with less suspicion than only a few years earlier and support from governments and OAU were obtained without difficulties. It might also have helped that the agency requesting import of parasitoids was itself a very potent donor.

The later successful parasitoid, *G. tebygi*, was quickly distributed in Togo from the insectary in Cacaveli (Agricola et al. 1989) and in other West and Central African countries from the IITA insectary in Cotonou (Neuenschwander 1989). In fact, collaboration in Nigeria was so good that IITA was invited to and officially thanked during a ceremonial release at the National Institute for Horticulture in 1989, a recognition not received for the vastly more intensive and important work in Nigeria with the exotic natural enemies of the CM.

As was the case with the CM, the pest status of the MM was so obvious that no quantitative crop loss assessment was done before the first releases. In fact, the relevant studies have only started now, in Benin. Favorable acceptance of the project was also facilitated because mango and ornamental trees, were especially hard hit in towns. Decision makers therefore often had a first-hand experience on their private properties, which had not often been the case with the CM.

Classical Biological Control Projects: Scientific Contributions
Cassava Mealybug

As is so often the case in biological control, foreign exploration for CM natural enemies in Central and South America resulted in an initial misidentification. This mealybug was later recognized as a new species, *P. herreni* Cox and Williams, whose parasitoids did not attack the intended host, *P. manihoti*. This unsuccessful foreign exploration, nevertheless led to a revision of the South-American mealybugs (D. J. Williams, unpubl. results). Had exploration stopped here, the project would have joined the long list of failed endeavors in biological control terminated too early, mostly for financial reasons.

The first *P. manihoti* from the New World was discovered serendipitously by an entomologist from CIAT in 1981 in Paraguay, an area previously earmarked for surveys by IITA. The same year, IIBC collected natural enemies on these CM populations. Intensive search over most cassava growing areas of South America finally yielded only eight areas in the Paraná River Basin, where the CM occurred (Löhr et al. 1990). CM populations most of the time were so low that no proper population dynamics studies were feasible and parasitoids had to be collected from artificially infested trap plants. This led to the collection of several hymenopterous parasitoids and coccinellid beetles, most of them still being reared in Cotonou.

Subsequently, natural enemy production at IITA was developed to the level where, by using local material wherever possible, actual production costs were similar to those for other mealybug parasitoids from commercial insectaries in Europe and North America (Neuenschwander and Haug 1992). This technology is now being transferred to the national programs. Since rearing of E. *lopezi* is initially difficult and handled very efficiently at IITA, the importance of local insectaries for the production of E. *lopezi* for releases is small. Most often, such insectaries serve instead purpose as centers of biological control for demonstration to visitors.

The releases of beneficials on about 150 occasions in Africa led to a spectacular result. By 1991, E. *lopezi* was established in 26 countries (most recently in Kenya) over an area of about 3 million km2, the coccinellids *Hyperaspis notata* (Mulsant) in the East African highlands, and *Diomus sp.* in Kinshasa (as shown in Figure 6.1). The most interesting fact is the coverage of E. *lopezi* across all ecological zones from the dry Sahel to the rainforest and into the highlands. It seems that every single release of E. *lopezi* led to establishment.

Monitoring establishment and spread demands relatively little effort and was generally well executed, most recently in Gabon (Boussienguet et al. 1991). Long-term monitoring (more than one year and at short intervals), however, was not achieved except in Nigeria (Hammond and Neuenschwander 1990) (as shown in Figure 6.2), Ghana (Cudjoe 1990), and Benin (W.N.O. Hammond and R. Allomasso, unpubl. results). All these studies showed the predominance of E. *lopezi* among the natural enemies of the CM, and generally low CM populations.

While many biological control projects stop at the level described here (mostly for lack of funds) the project against the CM continued, by assessing the impact of E. *lopezi* in more detail, employing all available techniques (Hodek et al. 1972, Kiritani and Dempster 1973, DeBach et al. 1976, van Lenteren 1980, Neuenschwander and Gutierrez 1989). The results from the long-term population dynamics studies (Figure 6.2) are probably the best demonstration of E. *lopezi's* impact. In addition, exclusion experiments, first done in the transition zone of Nigeria (Neuenschwander et al. 1986) and later successfully repeated in the rainforest zone of Ghana (Cudjoe et al. 1992) (as shown in Figure 6.3), showed the impact of this exotic parasitoid. A computer simulation model with three trophic levels, based on laboratory and small scale field trials, demonstrated the ten-fold reduction in CM population peaks achieved by E. *lopezi*, while coccinellids were responsible for a 25% reduction only, compared to the situation without coccinellids (Gutierrez et al. 1988a,b) (as shown in Figure 6.4).

Finally, surveys indicated that the full benefit of E. *lopezi* was achieved only two years after the release (Neuenschwander et al. 1992). Countrywide surveys in many countries, based on an unbiased choice of fields, thereby confirmed E. *lopezi's* efficiency. This was true also for the Congo (Hammond et al. 1989), where studies on a few chosen fields previously led

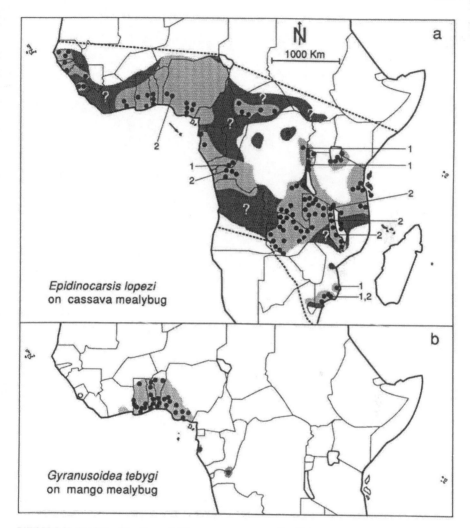

FIGURE 6.1a Distribution of *Phenacoccus manihoti* (dark) and
its introduced parasitoid *Epidinocarsis lopezi* (grey, dots =
release sites with establishment, arrow points to occurrence
of *E. lopezi* on Annobn Island), together with local
establishment of two exotic coccinellids, *Hyperaspis* sp. (1)
and *Diomus* sp. (2).

FIGURE 6.1b Same for *Rastrococcus invadens* and its
introduced parasitoid *Gyranusoidea tebygi* (circle indicates
recent release of *G. tebygi*). In addition, *Anagyrus mangicola*
was released but not yet recovered in central Benin, southern
Ghana, and Sierra Leone.

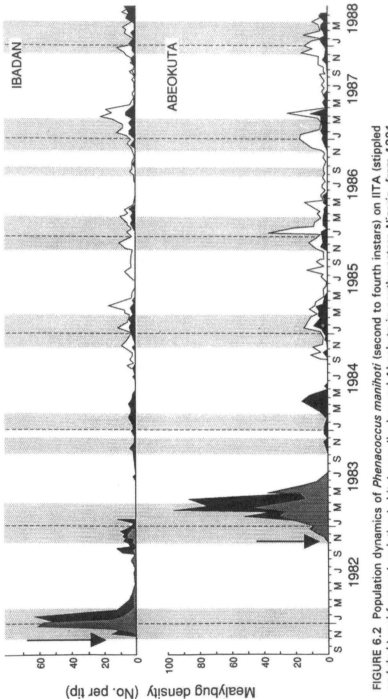

FIGURE 6.2 Population dynamics of *Phenacoccus manihoti* (second to fourth instars) on IITA (stippled and dark) and farmers' varieties (white) near Ibadan and Abeokuta in southwestern Nigeria, from 1981 to 1988. Arrows = releases of *Epidinocarsis lopezi* stippled = release fields, black = control fields, which were soon invaded by the spreading *E. lopezi*, vertical stippled bands = months with less than 60 mm rain (adapted from Hammond and Neuenschwander 1990).

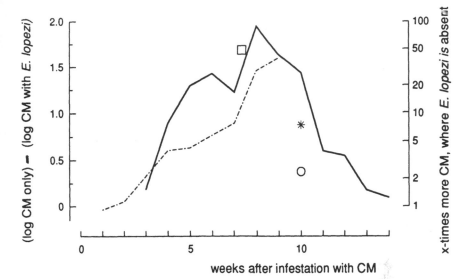

FIGURE 6.3 Difference in *Phenacoccus manihoti* (CM)
infestation between cassava plants from which *Epidinocarsis
lopezi* was excluded experimentally and those on which *E.
lopezi* was active (= control), expressed as multiple of the
control (or difference in log-transformed values). Lines =
chemical exclusion experiments in Nigeria (solid line) and
Ghana (broken line), points = physical exclusion for about 2
months in Nigeria (asterisk = late dry season, circle = early
dry season) and Ghana (square = middle of dry season). Data
for Nigeria from Neuenschwander et al. 1986, those for Ghana
from Cudjoe et al. 1992.

to the conclusion that *E. lopezi* was a failure (Le Rü et al. 1988). In these
and all previous IITA surveys, a small proportion of fields (about 5%), all
characterized by exceedingly bad soils, i. e. leached sandy soils in high rainfall areas and mostly cultivated repeatedly for many years, had unacceptably high CM populations. Under these conditions, mulching seemed to
offer an agronomic solution under the umbrella of biological control
(Neuenschwander et al. 1990). This conclusion is supported by fertilizer
experiments (Okeke 1990) and ties in with results on tillage and mulching
effects on cassava (Ohiri and Ezumah 1990, Ehui et al. 1991). This soil-
plant-CM-parasitoid interaction is now being investigated in more detail.

It was suggested that "reinforcing" releases with *E. lopezi* in these pockets of high infestation (Emehute and Egwuatu 1990) made no entomological sense, even if it is done before the population build-up. On these
largely defoliated plants with high CM populations, *E. lopezi* was shown
to be present, but no longer within the range of its density dependent reaction. Moreover it was observed that female wasps favor dense foliage. It

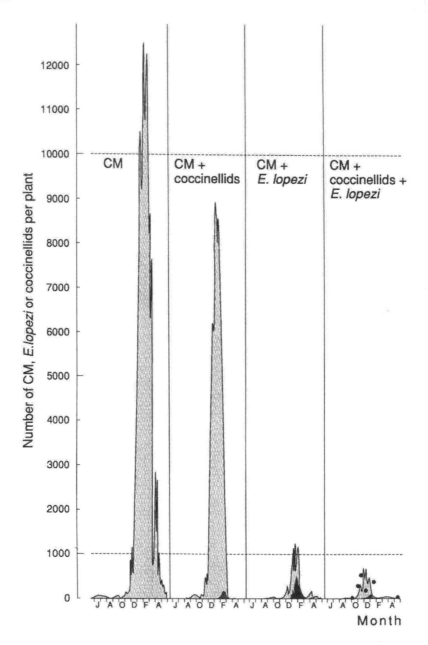

FIGURE 6.4 Population dynamics of *Phenacoccus manihoti* (CM) alone, with indigenous predators (coccinellids = dark grey), introduced parasitoid (*Epidinocarsis lopezi* = black), or both (black for *E. lopezi* covers grey for coccinellids), as predicted by a simulation model and validated by field data (dots) (adapted from Gutierrez et al. 1988a).

must be assumed that the relatively few females added in inundative releas es one per 10 tips on 10 ha. would already strain most insectaries, and would behave similarly to the already present wasps and no efficient control would ensue. Though "reinforcing" releases were made in a few instances, they must be viewed as a political palliative, serving for information and publicity only.

Overall, the cassava plant, CM, and *E. lopezi* proved to be the main factors influencing the pest insect. Studies concerning competitors and antagonists in the same food-web gave additional insights. Among the indigenous predators of the CM, coccinellids had been investigated extensively. Most studies ended with a vague recommendation to use these beetles in biological control of the CM, through inundative releases or by habitat management. In our view, these are applications that have their place in highly disturbed habitats on cash crops. In Africa, most cassava fields and their neighborhoods have abundant weeds that provide alternate food sources. Moreover, the economic situation of smallholders would hardly allow the purchase of natural enemies even if they were available. The contribution of the indigenous predators therefore seems largely immune to practical management and, as shown above, far inferior to the one of the adapted exotic parasitoid.

The biology of one indigenous hyperparasitoid has recently been described in detail (Goergen and Neuenschwander 1990), and as a group they have been studied extensively in the field. Due to the high mortality that these ubiquitous hyperparasitoids inflict on *E. lopezi* shortly after its establishment when CM and *E. lopezi* populations are very high, hyperparasitoids have drawn the attention of most national biological control programs. Hyperparasitoids were shown repeatedly to have a strong density dependent reaction to *E. lopezi*. Where *E. lopezi* becomes efficient in permanently reducing the populations of its host, *E. lopezi* itself becomes rather rare, whereupon the pressure by hyperparasitoids diminishes to low levels.

The establishment of *E. lopezi* also led to the competitive exclusion of the African *Anagyrus nyombae* Boussienguet (Hym., Encyrtidae), a rather inefficient parasitoid attacking the CM. Following the introduction of *E. lopezi*, it disappeared from the cassava based CM food web (Boussienguet et al. 1991). Later, this species was caught in yellow pans in the forest where it now seems to be restricted to its original, yet unknown host, which it attacked before the introduction of the CM.

In addition to *E. lopezi*, pathogens (Le Rü 1986, Le Rü et al. 1985, Le Rü and Iziquel 1990) and other exotic natural enemies of the CM are being studied as potential biological control agents. Food-web studies in South America on the two *Phenacoccus* spp. revealed one common parasitoid, *Epidinocarsis diversicornis* (Howard). This species is uniparental on the biparental *P. herreni* in northern South America, and biparental on the uniparental *P. manihoti* in southern South America. There was some hope that the biparental strain of *E. diversicornis* could be established in pockets of CM infestation. After several years of releases in 13 African countries, no long-term establishment was registered. Detailed laboratory and field investigations demonstrated why *E. diversicornis* was competitively displaced by the ubiquitous *E. lopezi* (J. J. M. van Alphen and P. Neuenschwander unpubl. results). It is inferior to *E. lopezi* in host finding and in direct larval competition, and it has a male biased sex-ratio on host instars on which *E. lopezi* produces a balanced sex-ratio. This system is now being used to explore, by simulation modeling, different release strategies for biological control projects with several natural enemy candidates (A. P. Gutierrez, J. J. M. van Alphen and P. Neuenschwander unpubl. results).

Overall, research of biological control of the CM yielded scientific contributions beyond the success of permanently suppressing, i.e., controlling, *P. manihoti*. Our results support the classical view with its central tenet of density dependent reaction of the natural enemy to its host populations (DeBach and Schlinger 1964, Huffaker et al. 1976, May and Hassell 1988). This density dependence is, however, only detectable if the sampling unit, in this case the tip, represents a unit that can be perceived by the ovipositing parasitoid female. If simple field means with their numerous zero counts are used, the counter-view that density dependence is not essential for biological control (Murdoch et al. 1985, Murdoch 1990) seems to prevail.

Host finding and selection (van Alphen and Vet 1986), sex allocation (Waage 1986, van Dijken et al. 1991), and aggregation (Hirose et al. 1990, Kareiva 1990, Hammond et al. 1991) were investigated not only in the laboratory but also in the field, making *E. lopezi* one of the best known encyrtid parasitoids. These studies reveal *E. lopezi* as a "single minded" insect with one great advantage over its competitors, namely its astounding searching capacity. Far from relaxing biological control over time through a possible genetic feed-back mechanism (Pimentel 1961), *E. lopezi* succeeds in keeping its potential competitors, coccinellids, out of the game by maintaining low host populations at a low level.

Mango Mealybug

Exploration for the MM was conducted mainly out of the IIBC station in India (Narasimham and Chacko 1988). Results on population dynamics were difficult to obtain because of the rarity of the MM and its occurrence together with several other *Rastrococcus* spp. Two specific parasitoids were processed by quarantine in London and sent to Togo, and from there to Benin.

Production of *G. tebygi* was based on relatively few data on its biology

(Willink and Moore 1988). Monitoring the MM, following the release of *G. tebygi* in Togo, quickly confirmed its establishment. Its capacity to spread seemed limited initially, but then reached distances comparable to those previously obtained with *E. lopezi* (Agricola et al. 1989). The indigenous hyperparasitoids attacking *G. tebygi* were the same species as those on *E. lopezi* (Agricola and Fischer 1991). Releases also led to establishment of *G. tebygi* in Benin, Nigeria, Gabon, Zaire. Without further releases, the parasitoid established itself in Congo and Côte d'Ivoire (Figure 6.1).

First results indicating the strong impact of *G. tebygi* were obtained in Togo (Agricola et al. 1989) and are being pursued in the other countries. Sampling procedures are difficult because of the complexity of the plant structure and the alternate bearing of mango trees (Singh 1968). The spatial distribution of the MM dictates that young and old leaves should be considered separately (Boavida et al. 1992). Apart from an overall reduction of the MM populations, a marked restriction of the host range of this initially very polyphagous mealybug has been observed, with mango trees often remaining the only hosts. In the country side, biological control of the MM by *G. tebygi* in most cases is excellent.

In some towns, however, damaging MM populations persist on mango trees and, occasionally, on frangipani. In these pockets, where sooty mold covers the leaves because of abundant honeydew production, a closely related encyrtid, *Anagyrus mangicola* Noyes, has been released in Benin, Ghana, and Sierra Leone in 1991. On the basis of a cursory investigation of the biology of this species (Cross 1990), a simulation model was developed and it was concluded that *A. mangicola* was not a promising species (Godfray and Waage 1991), a prediction that is now being tested in the field. In addition, pathogens have been described (Garcia and Moore 1988), but no field studies have yet been conducted.

Compared to many other biological control projects, the two projects on mealybugs in Africa stand out by their intensive research following an ecosystems approach and by the extensive scientific collaboration.

Clients and Beneficiaries: The Sociological Context

These scientific achievements were not reached in the proverbial ivory tower of research, but depended on important institutional and individual human interactions. To understand the impact of these results or sometimes the seeming lack of impact, we first discuss the different actors.

IITA, as all other centers of the Consultative Group on International Agricultural Research (CGIAR), is financed by donors, which in turn are subject to their national parliaments. The biological control program described above was financed by many of the same donors that support IITA, but mostly by special project funds or, lately, restricted core funding. The

CGIAR is advised by the Technical Advisory Committee (TAC) while the Biological Control Program, in addition, was advised by the EAC. Long-term plans for IITA were laid down in two strategic plans (IITA 1988a, b) and those for biological control in minor documents. Overall, oversight of our research activities was very intensive, with the aim of assuring that the clients benefitted from these activities and that this benefit was seen by the donors. Thus the activities of individual scientists, who work at IITA for a combination of reasons (material gain, career advancement, adventure, idealism), are weighed and shaped by outside forces to a much greater degree than is the case at universities, which do similar research.

IITA is not alone in the field of assisting the development of agriculture in Africa. Other international organizations like the International Center of Insect Physiology and Ecology (ICIPE), the Institut Francais de Recherche pour le Développement en Coopération (ORSTOM), FAO, IIBC, and some non-governmental organizations, some with a longer involvement in Africa than IITA, are potential competitors for donors' support. In some of these organizations, scientists are strongly committed to make their mark for an eventual career in their home country, an attitude which can stifle collaboration but may also bring a healthy portion of scientific competition.

One potential source for friction, commercial interests, did not play a role in these classical biological control projects on subsistence crops. Insecticide applications are largely out of the question and in most countries no market for cuttings or seedlings developed, even for farmers that could pay the price, because no attractive resistant varieties were available.

The stated clients of this research are the National Agricultural Research Systems (NARS), i.e. specialized institutions and extension services, including universities, which in their respective countries of course do not go by the name of NARS. These national institutions can provide the administrative framework, but the research still depends on gifted scientists and enlightened administrators, who let them execute it. Making this system work is already difficult enough in developed countries (Huffaker and Smith 1980, Bigler 1989). In Africa, the prime constraint is sustained "adequate financial, human and infrastructural resources in order to avoid further degradation of existing resources", as expressed in an IPM workshop in 1990 (Neuenschwander et al. 1991). Another important problem for the scientists in the NARS is the fact that they have to work as generalists, stretching themselves thin to cover a multitude of crops and pests. Universities in Africa face much the same problems.

The beneficiaries of the research should be the subsistence farmers. Ellis (1988) defines them as follows: "Peasants are farm households, with access to their means of livelihood in land, utilizing mainly family labor in farm production, always located in a larger economic system, but fundamentally characterized by partial engagement in markets which tend to function with a high degree of imperfection". Those with small participation in markets, subsistence farmers, mostly rely on traditional technology and often do not need IITA research. Population growth, however, often led to overutilization of limited resources straining traditional systems. This is the niche, where research by IITA can become useful. Large-scale farmers, on the other hand, usually have their own means to obtain information and are no longer seen as obligatory beneficiaries of IITA research. Overutilization of the land and the concomitant yield decline force farmers to migrate to the towns. The resulting, ever swelling ranks of the urban poor pose a special dilemma for IITA and other centers. Their clout is likely to grow in the new democracies arising in Africa, and their requirements are often opposed to those of the rural producers. Political decisions about keeping food prices low and the preference of short-term exploitation over long-term conservation are the consequences, both are opposed to an ideal rural development.

Working directly with the beneficiaries is not IITA's prime role, but where it is done, communication with farmers becomes the prime problem (Goodell et al. 1990). It is good to remember that the relationships among farmers are as infinitely complicated (Hill 1970) as those many researchers take for granted in their own societies. And looking at subsistence farmers in a romantic glow may be very satisfactory for the researcher, but will not bring any benefits to the farmer. Communication with colleagues from national institutions and farmers often hurts itself at the direction of the flow of information between different disciplines and between researchers at international centers and national research institutions, extensionists, and farmers (Goodell 1984, Weiss and Robb 1989, Goodell et al. 1990, Bentley and Andrews 1991, Pelletier and Msukwa 1990, Haverkort 1991). Because the common approach is top-down, much indigenous technical knowledge thereby risks to be lost, sometimes forever (Chambers and Jiggins 1986). In addition, problems might not receive the right priority because communication is disturbed due to different cultural attitudes and often gender (Jiggins 1990). To a minor degree the same is actually true for the flow of information between universities and donors in Europe and America and the CGIAR scientists, who live nearer to the beneficiaries.

Small scale farmers in Africa are politically often mute. This points to the importance of the press to feed back information about them to people in the north, who through their parliaments shape the activities of the donor agencies. Journalism, ideally at least, thus provides the feed-back in the information and decision making web.

In conclusion, the question concerning the impact of research by international centers on their clients has to take into account complicated human relationships. The impact arrow becomes more a food-web of mutual interactions with the dollar sign as main currency. The fact that each member has its own agenda, pre-programs conflicts. They can only be solved if the aim of a project is at least partly common to all participants and based on commonly agreed objective criteria.

Evaluating the Impact of Classical Biological Control

Problems in evaluating projects in rural development, including IPM, have been described vividly (Goodell et al. 1990, Elwert and Kretschmer 1991, Bentley and Andrews 1991). In the CM biological control project, the donors commissioned a study on crop loss that should go beyond the assessment of the CM's impact measured in rigorous field trials undertaken at IITA in 1982–83 (Schulthess et al. 1991). The study by international experts (Walker et al. 1982) confirmed the scientifically soft information obtained previously from governments and surveys by IITA entomologists, but it remained unsatisfactory to the EAC. It led, however, to a clarification of the techniques to be developed and, eventually, to a survey that quantified yield loss (Neuenschwander et al. 1989). Apart from this direct intervention from outside the project, continuous reviews by EAC assured donors and IITA administrators alike that the projects were following the right course. Collaboration with leading universities kept research from becoming parochial. The good number of publications in refereed journals, as cited in the last review (Herren and Neuenschwander 1991), attests to the successful scientific evaluation of biological control of the CM in all its aspects. A similar evaluation is underway for the MM biological control.

The sociological aspects of the two biological control projects, however, are evaluated for the first time here. Generally, interactions with the donors were very positive, resulting in the construction of the Biological Control Center for Africa and cumulating in the fact that the director of the program, Dr. H. R. Herren, was awarded the prestigious Rank Prize. It may be illustrative for future projects to note that one important potential donor was scared off early on by a well-meaning but highly ambiguous consultant report.

Throughout the project, contacts with the press were rather intensive, though sometimes with mixed results. Scientific and journalistic language often proved rather incompatible, leaving both sides frustrated. Especially in the beginning of the project, most newspaper texts did not report the given scientific information correctly. We can only guess that journalists found our way of thinking in small but concrete details equally difficult. In addition, two films were produced about the CM project and aired widely on television in Europe. How far this publicity and the numerous presentations by IITA staff influenced the donors cannot be determined, but we like to think that it helped.

In the beginning, the biological control project met, however, resistance and lack of understanding inside IITA. Outside support for the project eventually led to its acceptance and to the necessary administrative changes. This development cumulated in the creation at IITA of a Plant Health Management Division in 1992 around the already existing Biological Control Program, as discussed before.

As a brash newcomer on the African continent, the biological control project ruffled feathers also among internationally active institutions competing for funds. Eventually, collaborative projects were undertaken with all of the mentioned institutes, including CIAT and ICIPE, and have remained particularly strong with FAO, GTZ and IIBC, which have one staff each posted at the Biological Control Center for Africa in Cotonou. It can be surmised that IITA's biological control projects have not taken away donor funding from the other institutions, but actually heightened awareness of the potential benefits of this approach and therefore facilitated fund raising for biological control by other institutions in and outside Africa.

Generally, relationships with universities cannot be institutionalized the same way as those with international institutes. We collaborate very successfully and to mutual benefit with a few highly specialized university institutes in Europe and North America. The wished for broad based collaboration with African universities, however, has unfortunately not materialized. Practically all colleagues at African universities suffer from a degree of underfunding not known in the industrialized countries. Only a few colleagues could be financially supported and without financial support, collaboration was often difficult.

Any biological control project in Africa needs a special relationship with the IAPSC, headquartered in Yaound, Cameroon. After a stormy initial phase, the relationship is now very good and yearly visits assure good communication. We are no longer accused of releasing exotic organisms illegally, a reproach that is the bane of any biological control practitioner not working in his or her own country.

Due to the special relationship between biological control and quarantine authorities, the only clients of our project are the NARS. Nongovernment organizations wishing to receive beneficial insects always have to pass through the respective national quarantine authorities, and IITA provides insects only for official certificates. Since NARS are only viable if their scientists are dedicated and well educated, a massive program of training was undertaken by the UNDP/FAO sponsored training program, including training courses at IITA 2–3 times per year, in-country training courses, and degree-related training at African universities or overseas (Wodageneh, 1989). Thus IITA's research (basic and technology development) was linked with UNDP/FAO-sponsored training and GTZ-supported coordination of country programs financed by several donors, making an efficient formula for furthering rural development.

Despite these efforts, work between IITA's Biological Control Program and the NARS has a somewhat mixed review. The major positive impact was that there certainly is now a general awareness among African plant protection practitioners and policy makers of the potential benefits of biological control. Three stages in the quality of collaboration can be distinguished. Many country programs flourished because of individual scientists, who benefitted from our training program and received our direct support as well as help for bilateral arrangements with individual donors. In a few countries, collaboration with some individual scientists in national institutes is good, while other scientists in the same country seem to view IITA staff as intruders on their territory, so that jealousy and competition make collaboration with the national program difficult. Today, there is no longer any direct resistance to collaboration with IITA in any country. But in some countries, poor collaboration is observed because of continuously changing administrative set-ups, so that IITA can never deal with the same scientist for the extended period needed to complete a project.

Difficulties can also arise when extension entomologists bring highly infested cassava plants to the attention of policy makers without placing this observation into a larger context, i.e., without basing their conclusions on random samples. It can only be stressed that biological control, if effective, lowers the mean pest population density. Highly infested plants can still occur for purely statistical reasons. Where they are clustered, we try to find out the reasons and possible remedies, as discussed above. In one case, such biased information led to interesting discussions by an anthropologist about the role of information systems in combatting the mealy bug disaster (Pelletier and Msukwa 1990). These authors worked in areas that, during the second half of their field study and unknown to them, were already under good biological control (Neuenschwander et al. 1992). At about the same time, we experienced that all answers by farmers about the pest status of the CM became politically motivated, since farmers perceived foreigners asking questions to be linked to the free food aid, which had meanwhile started flowing.

It seems to be an illusion to assume that the interests of IITA's clients, the NARS, should always be the same as those of IITA, namely the presumed benefit to peasant farmers. On the side of the IITA scientist, mainly career advancement and often a necessity in view of the short-term contracts offered, can get in the way. On the side of the NARS, salaries that are insufficient for survival and lack of equipment and working facilities make it understandable why decisions are sometimes taken that do not gibe with any possible advancement for the stated beneficiary, the smallholder.

In the case of the two biological control programs, all growers of cassava and mango, smallholders, commercial farmers, and home owners alike, profited because these biological control projects were scale-neutral. Large-scale farmers and some urban gardeners certainly understood the projects. For the benefit of smallholders, publicity material in the form of posters and brochures were distributed to the extension services. Despite this, most farmers and some extensionists were not aware of the existence of biological control by *E. lopezi*, as shown in the following four examples. 1. Asked about the CM in Ghana and Côte d'Ivoire, farmers clearly recognized the drop in pest incidence in areas where *E. lopezi* had been present for the duration of the entire cassava crop, as opposed to those in areas where *E. lopezi* was still lacking. However, all farmers attributed this drop to weather conditions, an explanation that could be excluded, when weather data from different areas and years were compared (Neuenschwander et al. 1989). 2. When farmers in Nigeria were asked for which reasons they had changed varieties in the past 30 years, mealybugs were the second most important reason, after yield related considerations, between the 1970s and 1985. Before this period and since 1985, very few farmers listed the CM as a reason for changing varieties (Ay 1991). Since no varieties became abundant that would be resistant to CM, this answer shows the impact by *E. lopezi*, again unrecognized by the farmers. 3. From 1980 to 1985 cassava production in Oyo State doubled, while higher yielding new varieties accounted for about 25% of all cassava grown (Ikpi et al. 1989). Part of this increase is considered by us to be attributable to the effect of *E. lopezi*, which spread and became efficient in Oyo state during this period. 4. Similarly, interviews about the MM in villages, where pest populations had collapsed completely, frequently yielded the statement that the MM had "miraculously disappeared". So, despite an active publicity campaign, biological control was not generally recognized by the farmers as the reason for the observed reduction in yield loss. Clearly, a more extensive information campaign would have improved this situation, but a larger campaign could not be justified because of its high costs and its recognized lack to improve the biological control result itself.

A second category of potential beneficiaries, the urban poor, certainly gained from the biological control programs in that the commodities were accessible on the market. In Ghana, in part thanks to the activity of the newly established *E. lopezi*, prices of cassava fell again to the level found before the introduction of the CM. In Benin, following the wide-spread establishment of *G. tebygi*, mangos appeared on the market in 1990 for the first time in several years. The situation is different for those urban poor, who grow their own produce in backyards. Biological control of the CM, despite the presence of *E. lopezi*, is often not effective enough on these bad and repeatedly used soils to prevent CM damage. Similarly, damaging MM populations persist locally in large towns, though we have not yet unravelled the ecological reasons for this.

The often observed lack of understanding of the causes for a sudden decline in the pest population did not affect the course of the projects. However, once the exotic beneficials were established, their maintenance in the system was affected by insecticide mis-use, as in any IPM program. Thus, outbreaks of the CM were observed following wide-spread aerial applications of insecticides against the variegated grasshopper, *Zonocerus variegatus* L., in Ghana and Nigeria or near insecticide treated cotton fields in Benin. Insecticide mis-use is also likely on mangos in home gardens. For the benefit of future projects and for maintaining the benefits of biological control, a basic understanding of its working by the directly concerned smallholders could therefore be valuable in the future and merits further efforts in extension type activities.

Apart from some impressive extrapolations (a benefit-cost ratio of 149 to 1 was calculated by Norgaard (1988), but some basic assumptions have been questioned), economic analyses of the two projects are yet outstanding and are only recently undertaken. FAO statistical data on cassava, being the information the governments give FAO to publish, are sometimes unreliable or insufficient. Additional statistics on cassava and its use are therefore being collected across Africa by field evaluations and farmers' interviews in the so-called COSCA study (Nweke 1988, Carter and Jones 1989, Nweke et al. 1989 a, b). These studies take into account that cassava to a large extent is a commercial crop, hence prices reflect supply. In addition to the tons of tubers produced per hectare, the value of leaves and stems (Dahniya et al. 1981, Ezumah 1987) is also considered in the COSCA study.

For mangos, the economic evaluation is even more difficult than for cassava. Mangos are sold on most markets, but mango trees are also often planted in school yards and the fruits are picked and eaten by children, for whom they are important in nutrition (vitamins, etc.). Alternate bearing, widely observed unnecessary tree felling following the attack by the MM, and the general lack of official statistics further complicate the economic analysis, as it is now undertaken in Benin. In addition, such economic analyses should also consider that pests, and by extension their natural enemies, are common property (Feder and Regev 1975, Regev et al. 1976). Where the exotic parasitoid spreads fast, as was the case in the two mealybug studies, the influence of neighboring fields is minimal. However, in the case of a slow moving phytoseiid mite, like the one recently established against the cassava green mite, *Mononychellus tanajoa* (Bondar) (Acari, Tetranychidae), in Benin (Yaninek et al. 1992) the question of how the neighbor treats his fields becomes important.

Conclusion: The Contribution of Research in Classical Biological Control to Sustainable Agriculture

Classical biological control can be viewed as a large scale ecological experiment. Such research in applied ecology needs a vast amount of data, which are often either not gathered yet or lie fallow because of lack of capacity to analyze them (Slobodkin 1988). Despite the appealing set-up whereby insects and their offspring are permanently marked, i.e. the introduced species, classical biological control is often not taken seriously as academic research. At times the need for a scientific approach, with accurate measurements, repetitions, and disinterested analysis with the help of statistical methods, has ever been questioned (Norgaard 1987) and biological control has been likened more to an art than a science (van Lenteren 1980). As a result, classical biological control is sometimes considered staid and not at the forefront of science, though it is well accepted by the public and the donors. It is therefore relatively little represented in new books and proceedings, at the expense of more appealing, so-called modern, techniques. No wonder then that institutions dealing with biological control in universities are dwindling.

There are simply too few cases of successful biological control that are thoroughly investigated, though reviews listing the potential are common. It is in this rather gloomy context that the CM project takes on a particular importance. At IITA itself it has now led to further classical biological control projects (in collaboration with many of the previously mentioned institutions) against the cassava green mite, *M. tanajoa* (Yaninek et al. 1992), the cowpea thrips, *Megalurothrips sjostedti* (Tryb.) (Thys., Thripidae) (Tam 1991), the larger grain borer, *Prostephanus truncatus* (Horn) (Col., Bostrichidae) (Markham and Herren 1990), the banana weevil, *Cosmopolites sordidus* (Germar) (Col.:Curculionidae) and, most recently, the water hyacinth, *Eichhornia crassipes* (Martius) (Pontederiaceae). Most projects are new investigations with unknown potential, whereas water hyacinth biological control is well tested (Harley 1990). In addition, systems research in maize and cowpeas may turn up other possibilities for classical biological control.

We think that many more cases of pest insects or even failed biological control projects merit a second look and more intensive investigation. This might lead to a better match of original area and the pest species, or the discovery of cryptic species (Delucchi et al. 1976, Gauld 1986, Paterson 1990) with different host-parasitoid associations. This approach seems particularly promising for so-called circumtropical species, where ancient trade has obscured the origin of the species and where, in comparison to the large number of species present, relatively few have been investigated in detail. Investigating the potential for biological control asks, however, for more investment in foreign exploration and systematics. Both disciplines are woefully underfunded.

The big advantage of classical biological control as being safe, i.e., without environmental or health risk (Caltagirone and Huffaker 1980), has recently come under attack (Howarth 1991). Outside the field of biological control with vertebrates or molluscs, the introduction of which has been criticized repeatedly, very few examples have resulted in a loss of indigenous species. Classical biological control with relatively specific, insect parasitoids and predators, has not led to problems. Exceptions to this statement are only reported from extreme situations, where the natural environment, whose biodiversity was to be preserved, had shrunken to small remnants. It seems that classical biological control should better be judged against what would have happened without introduction of exotic species for biological control, than against a romantic notion of eden. There are many reasons why biodiversity should be maintained (Wilson and Peter 1988, Ehrlich and Wilson 1991), but in the case of classical biological control it can be argued that biodiversity was seriously impinged upon when the foreign pest insect was introduced. Thus, classical biological control in conditions where the equilibrium position of the pest populations after introduction of the exotic natural enemy is acceptable to the farmer, is the most ecologically advantageous solution to a pest problem. Of course, we are glad that the problem of extinction of an indigenous species never arose in the sometimes charged atmosphere, where *E. lopezi's* safety record seemed to cause more problems than that of the CM. Fortunately, *A. nyombae* was not eradicated.

Where classical biological control works, it is an economic solution (Mumford and Norton 1987). In fact, any proven small improvement, since it can accrue over infinite time, would lead to return of the costs. These returns can be judged against benefits from another possible investment of the same size as the cost of the biological control project. A good project with a high return, like the one with *E. lopezi*, can then carry the inevitable failures. The necessary studies on economic benefits in biological control are, however, rare (Dean et al. 1979) because the economy is often not clear due to unrecognized benefits to the community and a complicated structure of society.

Classical biological control is also a sustainable solution, a criterion that has recently become increasingly important at the expense of simple production criteria (Weil 1990, Harmsen 1990). Thus, in the decade following *E. lopezi's* introduction, no increase in equilibrium levels of the CM was observed. In most instances, *E. lopezi* maintains and defends its niche by maintaining low CM population levels.

As more and more projects, which seem to provide ecologically and economically acceptable solutions, fail because farmers do not adopt them, awareness increases about the value of adoptability (Chambers and Jiggins 1986, Norgaard 1988, Batie and Taylor 1989). Since classical biological control adopts itself to the conditions, it scores the highest points possible on this criterion.

All these advantages make classical biological control an ideal potential solution. For smallholders in tropical countries, the appeal is particularly strong for the following reasons. 1. Many species are not yet investigated. The potential for new successful matches between hosts and their natural enemies is therefore high. 2. Mis-use of insecticides, though sometimes severe, is mostly local. Potentially successful biological control is therefore less likely to be disrupted by chemical control of a key pest, as is so often the case in Europe or North America. 3. Alternate hosts and food sources are readily available. Biological control by indigenous or exotic natural enemies is therefore more likely to succeed than in the sometimes sterile agricultural environments, particularly in North America. 4. Tolerance for damage on subsistence crops is rather high because often no other alternative solutions except cultural control measures are available. As a consequence, aesthetic standards of the farm produce are not unreasonably high as is often the case in industrialized countries. 5. Finally, serious sociological difficulties at the farm level observed in many development projects are mostly avoided because adoptability of biological control is very good. It can be predicted that this technology will become ever more important, particularly in Africa, because increasing traffic favors introductions of insects and mites of foreign origins. On arrival they find the climate conducive for reproduction all year round and quarantine services generally weak.

As this review has shown abundantly, this does not mean that problems in human interactions are rare when biological control is being implemented. In conclusion, classical biological control is one of the few technologies that improve production by maintaining or increasing sustainability at no political and socio-economic costs to the farmers. It requires, however, intensive research and carefully coordinated implementation, demanding a strong effort for mutual understanding.

Acknowledgments

The two projects were mainly financed by the aid agencies of Austria, Belgium, Canada (IDRC), Denmark, Federal Republic of Germany (GTZ), Italy, the Netherlands, Norway, and Switzerland, and the African Development Bank, the European Economic Community, the International Fund for Agricultural Development, as well as UNDP. I also thank my colleagues at IITA for stimulating discussions.

Bibliography

Agounké, D., Agricola, U. and Bokonon-Ganta, H. A. 1988. *Rastrococcus invadens* Williams (Hemiptera: Pseudococcidae), a serious exotic pest of fruit trees and other plants in West Africa. *Bull. Ent. Res.* 78: 695–702.

Agricola, U., Agounké, D., Fischer, H. U. and Moore, D. 1989. The control of

Rastrococcus invadens Williams (Hemiptera: Pseudococcidae) in Togo by the introduction of *Gyranusoidea tebygi* Noyes (Hymenoptera: Encyrtidae). *Bull. Ent. Res.* 79: 671–678.

Agricola, U. and Fischer, H. U. 1991. Hyperparasitism in two newly introduced parasitoids, *Epidinocarsis lopezi* and *Gyranusoidea tebygi* (Hymenoptera: Encyrtidae) after their establishment in Togo. *Bull. Ent. Res.* 81: 127–132.

Akobundu, I. O. 1990. "The role of weed control in integrated pest management for tropical root and tuber crops," in *Integrated Pest Management for Tropical Root and Tuber Crops.* S. K. Hahn and F. E. Caveness (eds.), : IITA, Ibadan. pp. 23–29.

Altieri, M. A. 1987. *Agroecology: The Scientific Basis of Alternative Agriculture.* Westview Press, Boulder. 227 p.

Altieri, M. A. and Hecht, S. B. (eds.). 1990. *Agroecology and Small Farm Development.* CRC Press, Boca Raton. 262 p.

Andrews, K. L. and Bentley, J. W. 1990. IPM and resource-poor central American farmers. *Global* 1: 6–9.

Arnold, M. A., Aeschlimann, J.P., Coaker, T. H., Murphy, H. T., Norgaard, R. B. and Rachie, K. O. 1987. *Midterm Review Africa-wide Biological Control Program of the International Institute of Tropical Agriculture.* Petit Jean Mountain, Morrilton. Winrock Internat., Arkansas. 96 p.

Ay, P. 1991. Spread and impact of cassava varieties in western Nigeria. Research report, unpubl. draft for *RCMP Res. Monograph.* IITA, Ibadan.

Batie, S. S. and Taylor, D. B. 1989. Widespread adoption of non-conventional agriculture: Profitability and impacts. *Amer. J. Alternative Agric.* 4: 128–134.

Beck, B. D. A. 1980. "Historical perspectives of cassava breeding in Africa," in *Root Crops in Eastern Africa.* Proc. Workshop, Kigali, Rwanda, 23–27 Nov. 1980. IDRC, Ottawa. pp. 13–18

Bentley, J. W. and Andrews, K. L. 1991. Pests, peasants, and publications: Anthropological and entomological views of an integrated pest management program for small-scale Honduran farmers. *Human Organization* 50: 1–23.

Bigler, F. 1989. Gedanken zum ntzlingsschonenden Einsatz von Pestiziden im integrierten Pflanzenschutz. *Schweiz. Landw. Fo.* 28: 49–55.

Boavida, M. C., Neuenschwander, P. and Schulthess, F. 1992. Spatial distribution of *Rastrococcus invadens* Williams (Hom., Pseudococcidae) in mango trees. *J. Appl. Ent.* (in press).

Boussienguet, J., Neuenschwander, P. and Herren, H. R. 1991. Essais de lutte biologique contre la cochenille du manioc au Gabon: I. Etablissement, dispersion du parasite exotique *Epidinocarsis lopezi* (Hym.: Encyrtidae) et déplacement compétitif des parasites indigênes. *Entomophaga* 36: 455–469.

Caltagirone, L. E. and Huffaker, C. B. 1980. Benefits and risks of using predators and parasites for controlling insects. *Ecol. Bull.* 31: 103–109.

Carson, R. 1962. *Silent Spring.* Houghton Mifflin Co, Boston.

Carter, S. E. and Jones, P. G. 1989. COSCA site selection procedure. *COSCA working Paper* 2: 19. IITA, Ibadan.

Chaboussou, F. 1970. Influence des pesticides sur la plante: conséquences écologiques. *Bull. Soc. Ecologie* 3: 146–158.

Chambers, R. and Jiggins, J. 1986. "Agricultural research for resource poor farmers: a parsimonious paradigm." IDS Discussion Paper 220. 38 p.

Croizat, L., Nelson, G. and Rosen, D. E. 1974. Centers of origin and related concepts. *Systematic Zool.* 23: 265–287.

Croft, B. A., Adkisson, P. L., Sutherst, R. W. and Simmons, G. A. 1984. "Applications of ecology for better pest control," in *Ecological Entomology.* C. B. Huffaker and R. L. Rabb (eds.). John Wiley and Sons, New York. pp. 763–795

Cross, A. E. 1990. "Studies of the ovipositional behavior and development of the immature stages of *Anagyrus mangicola* (Hymenoptera: Encyrtidae), a parasitoid of the fruit tree mealybug *Rastrococcus invadens* Williams (Hemiptera: Pseudococcidae)." M.S. Thesis. Birkbeck College, Univ. of London. 74 p.

Cudjoe, A. R. 1990. "Biocontrol of cassava mealybug in the rainforest zone of Ghana." Ph.D. Thesis. Wye College, London. 225 p.

Cudjoe, A. R., Neuenschwander, P. and Copland, M. J. W. 1992. Experimental determination of the efficiency of indigenous and exotic natural enemies of the cassava mealybug, *Phenacoccus manihoti* (Hom., Pseudococcidae), in Ghana. *J. Appl. Ent.* (in press).

Dahniya, M. T., Oputa, C. O. and Hahn, S. K. 1981. Effects of harvesting frequency on leaf and root yields of cassava. *Exp. Agric.* 17: 91–95.

Dean, H. A., Schuster, M. Boling, J. C. and Riherd, P. T. 1979. Complete biological control of *Antonina graminis* in Texas with *Neodusmetia sangwani* (a classic example). *Ent. Soc. Amer. Bull.* 25: 262–267.

De Bach, P. 1951. The necessity for an ecological approach to pest control on citrus in California. *J. Econ. Ent.* 44: 443–447.

De Bach, P. and Schlinger, E. I. (eds.). 1964. *Biological Control of Insect Pests and Weeds.* Chapman and Hall, London. 844 p.

De Bach, P., Huffaker, C. B. and MacPhee, A. W. 1976. "Evaluation of the impact of natural enemies," in *Theory and Practice of Biological Control.* C. B. Huffaker and P. S. Messenger (eds.). Academic Press, New York. pp. 255–285

Delucchi, V. 1987. "La protection intégrée des cultures," in *Integrated Pest Management—Protection Intégrée: Quo vadis?* V. Delucchi (ed.). Parasitis, Geneva. pp. 7–22.

Delucchi, V. 1989. "Integrated pest management vs systems management," in *Biological Control: A Sustainable Solution to Crop Pest Problems in Africa.* J. S. Yaninek and H. R. Herren (eds.). IITA, Ibadan. pp. 51–67

Delucchi, V., Rosen, D. and Schlinger, E. I. 1976. "Relationship of systematics to biological control," in *Theory and Practice of Biological Control.* C. B. Huffaker and P. S. Messenger (eds.). Academic Press, New York. pp. 81–91

Doutt, R. L. and Nakata, J. 1965. Overwintering refuge of *Anagrus epos* (Hymenoptera: Mymaridae). *J. Econ. Entomol.* 58: 586.

Edwards, C. A. 1989. The importance of integration in sustainable agricultural systems. *Agric. Ecosyst. Envir.* 27: 25–35.

Ehrlich, P. R. and Wilson, E. O. 1991. Biodiversity studies: science and policy. *Science* 253: 758–762.

Ellis, F. 1988. *Peasant Economics. Farm Households and Agrarian Development.* Cambridge Univ. Press, Cambridge. 257 p.

Elwert, G. and Elwert-Kretschmer, K. 1991. Mit den Augen der Beniner. Eine andere Evaluation von 25 Jahren DED in Bénin. Unpubl. report, 80 pp. and abbreviated in *Frankfurter Rundschau* 186: 13.

Ehui, S. K., Kang, B. T. and Spencer, D. S. C. 1991. Economic analysis of soil erosion effects in alley cropping, no-till, and bush fallow systems in southwestern Nigeria. *IITA Res.* 37: 1–6.

Emehute, J. K. U. and Egwuatu, R. I. 1990. Effects of field populations of cassava mealybug, *Phenacoccus manihoti*, on cassava yield and *Epidinocarsis lopezi* at different planting dates in Nigeria. *Trop. Pest Manage.* 36: 279–281.

Ezumah, H. C. 1987. The effect of harvesting leaves on cassava yield in Zaire. *Agric. Intern.* 39: 152–155.

Feder, G. and Regev, U. 1975. Biological interactions and environmental effects in the economics of pest control. *J. Envir. Econ. Manage.* 2: 75–91.

Gahukar. R. T. 1991. Current approaches to cassava pest management in subtropical Africa. *Agric. Zool. Rev.* 4: 97–136.

Garcia, E. F. and Moore, D. 1988. Hirsutella sphaerospora as a potential biocontrol agent of *Rastrococcus invadens*. Williams. *Proc. Brighton Crop Prot. Conf.-Pest and Diseases* 9C–2: 1125–1130.

Garcia, R., Caltagirone, L. E. and Gutierrez, A. P. 1988. Comments on a redefinition of biological control. *BioScience* 38: 692–694.

Gauld, I. D. 1986. "Taxonomy, its limitations and its role in understanding parasitoid biology," in *Insect Parasitoids*. J. Waage and D. Greathead (eds.). Academic Press, London. pp. 1–21.

Gips, T. 1987. "Breaking the pesticide habit. Alternatives to 12 hazardous pesticides." Minnesota: IASA Publ. No. 1987–1. 372 p.

Goergen, G. and Neuenschwander, P. 1990. Biology of *Prochiloneurus insolitus* (Alam) (Hymenoptera, Encyrtidae), a hyperparasitoid on mealybugs (Homoptera, Pseudococcidae): immature morphology, host acceptance and host range in West Africa. *Bull. Soc. Ent. Suisse* 63: 317–326.

Godfray, H. C. J. and Waage, J. K. 1991. Predictive modelling in biological control: the mango mealybug (*Rastrococcus invadens*) and its parasitoids. *J. Appl. Ecol.* 28: 434–453.

Goodell, G. 1984. Challenges to international pest management research and extension in the third world: Do we really want IPM to work? *Bull. Ent. Soc. Amer.* Fall 1984: 18–26.

Goodell, G., Andrews, K. L. and Lôpez, J. I. 1990. The contribution of agronomo-anthropologists to on-farm research and extension in integrated pest management. *Agric. Systems* 32: 321–340.

Greany, P. D., Vinson, S. B. and Lewis, W. J. 1984. Insect parasitoids: Finding new opportunities for biological control. *BioScience* 34: 690–696.

Greathead, D. J. 1989a. "Prospects for natural enemies in combination with pesticides." Presented, FFTC-NARC International Seminar, The use of parasitoids and predators to control agricultural pests. Tukuba Science City, Japan, 2–7 Oct. 1989.

Greathead, D. J. 1989b. "Present possibilities for biological control of insect pests and weeds in tropical Africa." in *Biological Control: A Sustainable Solution to Crop Pest Problems in Africa*. J. S. Yaninek and H. R. Herren (eds.). IITA, Ibadan. pp. 173–194

Greathead, D. J., Lionnet, J. F. G., Lodos, N. and Whellan, J. A. 1971. *A Review of Biological Control in the Ethiopian Region*. Commonwealth Agri. Bureaux, Slough. 162 p.

Greathead, D. J. and Waage, J. K. 1983. "Opportunities for biological control of agricultural pests in developing countries." World Bank Tech. Paper 11. 44 p.

Gutierrez, A. P., Neuenschwander, P., Schulthess, F., Herren, H. R., Baumgürtner, J. U., Wermelinger, B., Lhr, B. and Ellis, C. K. 1988a. Analysis of biological control of cassava pests in Africa. II. Cassava mealybug *Phenacoccus manihoti*. *J. Appl. Ecol.* 2: 921–940.

Gutierrez, A. P., Wermelinger, B. Schulthess, F., Baumgürtner, J. U., Herren, H. R. and Ellis, C. K. 1988b. Analysis of biological control of cassava pests in Africa. I. Simulation of carbon, nitrogen and water dynamics in cassava. *J. Appl. Ecol.* 25: 901–920.

Hagen K. S. 1976. Role of nutrition in insect management. *Proc. Tall Timbers Conf. Ecol. Animal Control by Habitat Manage.* 6: 221–261.

Hagen K. S. 1986. "Ecosystem analysis: plant cultivars (HPR), entomophagous species and food supplements," in *Interactions of Plant Resistance and Parasitoids and Predators of Insects.* D. J. Boethel and Eikenbary, R. D. (eds.). Ellis Horwood, New York. pp. 151–197

Hahn, S. K., Terry, E. R., Leuschner, K., Akobundu, I. O., Okali, C. and Lal, R. 1979. *Cassava improvement in Africa.* Field Crops Res. 2: 193–226.

Hahn, S. K., Isoba, J. C. G. and Ikotun, T. 1989. Resistance breeding in root and tuber crops at the International Institute of Tropical Agriculture (IITA), Ibadan, Nigeria. *Crop Protection* 8: 147–168.

Hammond, W. N. O. and Neuenschwander, P. 1990. Sustained biological control of the cassava mealybug *Phenacoccus manihoti* (Hom.: Pseudococcidae) by *Epidinocarsis lopezi* (Hym.: Encyrtidae). *Entomophaga* 3: 515–526.

Hammond, W. N. O., Nsiama She, H. D., Boussienguet, J., Ganga, T. and Reyd, G. 1989. Status of biological control of the cassava mealybug, *Phenacoccus manihoti* Mat.-Ferr. (Hom., Pseudococcidae) under various ecological conditions in Central Africa. *Proc. 4th. Triennial Symp. Inter. Soc. Trop. Root Crops, Africa Branch*, Kinshasa, 1989. IDRC, Ohowa. (in press).

Hammond, W. N. O., van Alphen, J. J. M., Neuenschwander, P. and van Dijken, M. J. 1991. Aggregation by field populations of *Epidinocarsis lopezi* (De Santis) (Hym.: Encyrtidae) a parasitoid of the cassava mealybug *Phenacoccus manihoti* Mat.-Ferr. (Hom.: Pseudococcidae). *Oecologia* (in press).

Harley, K. L. S. 1990. The role of biological control in the management of water hyacinth, *Eichhornia crassipes.* Biocontrol News Info. 1: 11–22.

Harmsen, R. 1990. The theory of sustainable agriculture: opportunities and problems. *Proc. Ent. Soc. Ontario* 121: 13–24.

Haverkort, B. 1991. Farmers' experiments and participatory technology development," in *Joining Farmers' Experiments. Experience in Participatory Technology Development.* B. Haverkort, J. van der Kemp and A. Waters-Bayer (eds.). ILEIA. Readings in Sustainable Agriculture. Intermediate Technol. Pub., London. pp. 3–16.

Hecht, S. B. 1987. "The evolution of agroecological thought," in *Agroecology: The Scientific Basis of Alternative Agriculture.* M. A. Altieri, (ed.). Westview Press, Boulder. pp. 1–20.

Heinrichs, E. A. 1988. *Plant Stress-insect Interactions.* John Wiley and Sons, New York. 492 p.

Herren, H. R. 1990. Biological control as the primary option in sustainable pest management: the cassava pest project. *Bull. Soc. Ent. Suisse* 63: 405–413.

Herren, H. R. and Neuenschwander, P. 1991. Biological control of cassava pests in Africa. *Annu. Rev. Entomol.* 3: 257–283.

Hill, P. 1970. *Studies in Rural Capitalism in West Africa.* Cambridge Univ. Press, Cambridge. 173 p.

Hirose, Y., Nakamura, T. and Tagaki, M. 1989. "Successful biological control: a case study of parasitoid aggregation." in *Critical Issues in Biological Control.* M. Mackauer, L. E. Ehler and J. Roland (eds.). Intercept Ltd., Andover. pp. 171–183

Hodek, I., Hagen, K. S. and van Emden, H. F. 1972. "Methods for studying effectiveness of natural enemies," in *Aphid Technology.* H. van Emden (ed.). Academic Press, London. pp. 147–188.

Hokkanen, H. and Pimentel, D. 1984. New approach for selecting biological control agents. *Can. Ent* 116: 1109–1121.

Hoy, M. A. 1985. Recent advances in genetics and genetic improvement of the phytoseiidae. *Annu. Rev. Entomol.* 30: 345–370.

Hoy, M. A. 1988. Biological control of arthropod pests: traditional and emerging technologies. *Amer. J. Alternative Agric.* 3: 63–68.

Howarth, F. G. 1991. Environmental impacts of classical biological control. *Annu. Rev. Entomol.* 36: 485–509.

Hueth, D. and Regev, U. 1974. Optimal agricultural pest management with increasing pest resistance. *Am. J. Agr. Econ.* 56: 543–552.

Huffaker, C. B. 1974. Some implications of plant-arthropod and higher-level arthropod-arthropod food links. *Envir. Entomol.* 3: 1–9.

Huffaker, C. B. (ed.). 1980. *Theory and Practice of Biological Control.* Academic Press, New York. 500 p.

Huffaker, C. B., Berryman, A. A. and Laing, J. E. 1984. "Natural control of insect populations," in *Ecological Entomology.* C. B. Huffaker and R. L. Rabb (eds.), John Wiley and Sons, New York. pp. 359–398.

Huffaker, C. B., Simmonds, F. J. and Laing, J. E. 1976. "The theoretical and empirical basis of biological control," in *Theory and Practice of Biological Control.* C. B. Huffaker and P. S. Messenger (eds.). Academic Press, New York. pp. 42–78.

Huffaker, C. B. and R. F. Smith. 1980. "Rationale, organization, and development of a national integrated pest management project,"in *New Technology of Pest Control.* C. B. Huffaker (ed.) Wiley and Sons, New York. pp. 1–24

Hussey, N. W. 1990. Agricultural production in the third world - a challenge for natural pest control. *Expl. Agric.* 26: 171–183.

IITA. 1988a. "IITA strategic plan 1989–2000." IITA, Ibadan.

IITA. 1988b. "IITA medium term plan 1989–1993." IITA, Ibadan.

Ikpi, A. E., Gebremeskel, T., Ezumah, H. C. and Ekpere, J.A. 1989. Cassava production in Oyo state, " in *Cassava: Lifeline for the Rural Household.* A. E. Ikpi and N. D. Hahn (eds.) Book Builders, Ibadan. pp. 39–59.

Jiggins, J. 1990. "Gender issues and agricultural technology development," in *Agroecology and Small Farm Development.* M. A. Altieri and S. B. Hecht (eds.). CRC Press, Boca Raton. pp. 45–51.

Kareiva, P. 1990. "The spatial dimension in pest-enemy interactions.," in *Critical Issues in Biological Control*. M. Mackauer, L. E. Ehler and J. Roland (eds.). Intercept Ltd., Andover. pp. 213–227.

Kennedy, J. S. 1977. "Olfactory responses to distant plants and other odor sources," in *Chemical Control of Insect Behavior. Theory and Application*. H. H. Shorey and J. J. McKelvey (eds.) John Wiley and Sons, New York. pp. 67–91.

Kiritani, K. and Dempster, J. P. 1973. Different approaches to the quantitative evaluation of natural enemies. *J. Appl. Ecol.* 10: 323–330.

Kirkby, R. A. 1990. "The ecology of traditional agroecosystems in Africa," in *Agroecology and Small Farm Development*. M. A. Altieri and S. B. Hecht (eds.). CRC Press, Boca Raton. pp. 173–180.

Kiss, A. and Meerman, F. 1991. "Integrated pest management and African agriculture." World Bank Tech. Paper 142. The World Bank, Washington. 122p.

Lal, R. 1982. "Effective conservation farming systems for the humid tropics" in *Soil Conservation in the Tropics*. Amer. Soc. Agron. and Soil Sci. Soc. Amer., Madison. pp. 57–76

Le Rü, B. 1986. Etude de l'évolution d'une mycose *Neozygites fumosa* (Zygomycetes, Entomophthorales) dans une population de la cochenille du manioc *Phenacoccus manihoti* (Hom.: Pseudococcidae). *Entomophaga* 31: 79–89.

Le Rü, B. and Iziquel, Y. 1990. Nouvelles données sur le déroulement de la mycose â *Neozygites fumosa* sur la cochenille du manioc *Phenacoccus manihoti*. *Entomophaga* 35: 173–183.

Le Rü, B., Iziquel, Y., Biassangama, A. and Kiyindou, A. 1988. "Comparaison des effectifs de la cochenille du manioc *Phenacoccus manihoti* avant et après introduction d'*Epidinocarsis lopezi*, Encyrtidae américain, au Congo," in *La Cochenille du Manioc et sa Biocénose au Congo: 1985–1987*. Travaux de l'équipe Franco-congolaise-ORSTOM-DGRST. ORSTOM., Paris. pp. 1–12

Le Rü, B., Silvie, P. and Papierok, B. 1985. L'entomophthorale *Neozygites fumosa*, pathogêne de la cochenille du manioc, *Phenacoccus manihoti* (Hom.: Pseudococcidae), en République Populaire du Congo. *Entomophaga* 30: 23–29.

Löhr, B., Varela, A. M. and Santos, B. 1990. Exploration for natural enemies of the cassava mealybug, *Phenacoccus manihoti* (Homoptera: Pseudococcidae), in South America for the biological control of this introduced pest in Africa. *Bull. Ent. Res.* 80: 417–425.

Markham, R. H. and Herren, H.R. (eds.). 1990. *Biological Control of Larger Grain Borer*. IITA, Ibadan. 171 p.

May, R. M. and Hassell, M. P. 1988. Population dynamics and biological control. *Phil. Trans. R. Soc. Bull.* 318: 129–169.

McMurtry, J. A. and Rodriguez, J. G. 1987. "Nutritional ecology of phytoseiid mites," in *Nutritional Ecology of Insects, Mites, Spiders, and Related Invertebrates* F. Slansky and J. G. Rodriguez (eds.) John Wiley and Sons, New York. pp. 609–644.

Michler, W. 1991. *Weissbuch Afrika*. J. H. W. Dietz Nachf. GmbH., Bonn. 568 p.

Mumford, J. D. and Norton, G. A. 1987. "Economic aspects of integrated pest management," in *Integrated Pest Management - Protection Intégrée: Quo vadis?* V. Delucchi (ed.). Parasitis, Geneva. pp. 397–407.

Murdoch, W. W. 1990. "The relevance of pest-enemy models to biological control," in *Critical Issues in Biocontrol*. M. Mackauer, L. E. Ehler and J. Roland (eds.), Intercept Ltd., Andover. pp. 1–24

Murdoch, W. W., Chesson, J. and Chesson, P. L. 1985. Biological control in theory and practice. *Amer. Nat.* 125: 344–366.

Narasimham, A. U. and Chacko, M. J. 1988. *Rastrococcus* spp. (Hemiptera: Pseudococcidae) and their natural enemies in India as potential biocontrol agents for *R. invadens* Williams. *Bull. Ent. Res.* 78: 703–708.

Natural Resources Institute (NRI). 1991. *A Synopsis of Integrated Pest Management in Developing Countries in the Tropics*. Natural Resources, Chatham. 20 p.

————. 1991. *A Synopsis of Pest Management in Developing Countries in the Tropics*. Natural Resources Institute, Chatham. 20 p.

Neuenschwander, P. 1989. Biocontrol of mango mealybug. *IITA Res. Briefs* 9: 5–6.

Neuenschwander, P., Borowka, R., Phiri, G., Hammans, H., Nyirenda, S., Kapeya, E. H. and Gadabu, A. 1992. Biological control of the cassava mealybug *Phenacoccus manihoti* (Hom., Pseudococcidae) by *Epidinocarsis lopezi* (Hym., Encyrtidae) in Malawi. *Biocontrol Sci. Technol.* (in press).

Neuenschwander, P. and Gutierrez, A. P. 1989. "Evaluating the impact of biological control measures," in *Biological Control: A Sustainable Solution to Crop Pest Problems in Africa*. J. S. Yaninek and H. R. Herren (eds.). IITA, Ibadan pp. 147–155

Neuenschwander, P., Hammond, W. N. O., Ajuonu, O., Gado, A., Echendu, N., Bokonon-Ganta, A. H., Allomasso, R. and Okon, I. 1990. Biological control of the cassava mealybug, *Phenacoccus manihoti* (Hom., Pseudococcidae), by *Epidinocarsis lopezi* (Hym., Encyrtidae) in West Africa, as influenced by climate and soil. *Agric. Ecosyst. Environ.* 32: 39–55.

Neuenschwander, P., Hammond, W. N. O., Gutierrez, A. P., Cudjoe, A. R., Adjakloe, R., Bäumgartner, J. U. and Regev, U. 1989. Impact assessment of the biological control of the cassava mealybug, *Phenacoccus manihoti* Matile-Ferrero (Hemiptera: Pseudococcidae), by the introduced parasitoid *Epidinocarsis lopezi* (De Santis) (Hymenoptera: Encyrtidae). *Bull. Ent. Res.* 79: 579–594.

Neuenschwander, P. and Haug, T. 1992. "New technologies for rearing *Epidinocarsis lopezi* (Hym., Encyrtidae), a biological control agent against the cassava mealybug *Phenacoccus manihoti* (Hom., Pseudococcidae)." in *Advances and Applications in Insect Rearing*. T. A. Anderson and N. Leppla (eds.). Westview, Boulder. (in press).

Neuenschwander, P., Herren, H. R. and Wodageneh, A. (eds.). 1991. *Integrated pest management in root and tuber crops. Protection intégrée des plantes â racines et tubercules*. IITA, Ibadan. 118 p.

Neuenschwander, P., Schulthess, F. and Madojemu, E. 1986. Experimental evaluation of the efficiency of *Epidinocarsis lopezi*, a parasitoid introduced into Africa against the cassava mealybug, *Phenacoccus manihoti Ent. Exp. Appl.* 42: 133–138.

Norgaard, R. B. 1987. "The epistomological basis of agroecology," in *Agroecology and Small Development*. M. A. Altieri and S. B. Hecht (eds.). CRC Press, Inc., Boca Raton. pp. 21–27

————. 1988. The biological control of cassava mealybug in Africa. *Amer. J. Agric. Econ.* 70: 366–371.

Nwanze, K. F. and Leuschner, K. (eds.). 1978. "Proceedings of the international workshop of the cassava mealybug *Phenacoccus manihoti* Mat.-Ferr. (Pseudococcidae). IITA, Ibadan. 85 p.

Nweke, F. I. 1988. COSCA project description. *COSCA Working Paper* 1: 31 IITA, Ibadan.

Nweke, F. I., Lynam, J. and Prudencio, C. (eds.). 1989a. Status of data on cassava in major producing countries of Africa (Cameroon, Cte d'Ivoire, Ghana, Nigeria, Tanzania, Uganda and Zaire). *COSCA Working Paper* 3: 35. IITA, Ibadan.

————. 1989b. Methodologies and data requirements for cassava systems study in Africa. *COSCA Working Paper* 4: 46. IITA, Ibadan.

Ohiri, A. C. and Ezumah, H. C. 1990. Tillage effects on cassava (*Manihot esculenta*) production and some soil properties. *Soil and Tillage Res.* 17: 221–229.

Okeke, J. E. 1990. "Status of the cultural management component in an integrated control of the cassava mealybug (*Phenacoccus manihoti* Mat.-Ferr.) and green spider mite (*Mononychellus tanajoa* Bondar) in Nigeria," in *Integrated Pest Management for Tropical Root and Tuber Crops*. S. K. Hahn and F. E. Caveness (eds.). IITA, Ibadan. pp. 188–192.

Okigbo, B. N. 1989. "New crops for food and industry: the roots and tubers in tropical Africa," in *New Crops for Food and Industry*. G. E. Wickens, N. Itaq, and P. Day (eds.). Chapman and Hall, London. pp. 123–134

Oldfield, M. L. and Alcorn, J. B. 1987. *Conservation of traditional agroecosystems*. BioScience 37: 199–208.

Organization of African Unity (O. A. U.). 1988. "Scientific, Technical and Research Commission, Interafrican Phytosanitary Council 1980," in *Interafrican Phytosanitary and Coordinated Regulation*. Vol 1. SOPECAM, Yaoundé. 207 p.

Paterson, H. E. H. 1990. "The recognition of cryptic species among economically important insects," in *Heliothis: Research Methods and Prospects*. M. P. Zalucki (ed.). Springer, Stuttgart.

Pelletier, D. L. and Msukwa, L. A. H. 1990. The role of information systems in decision-making following disasters: Lessons from the mealybug disaster in northern Malawi. *Human Organization* 49: 245–254.

Pickett, J. A. 1988. Integrating use of beneficial organisms with chemical crop protection. *Phil. Trans. R. Soc. London B.* 318: 203–211.

Pimentel, D. 1961. On a genetic feed-back mechanism regulating populations of herbivores, parasites and predators. *Amer. Nat.* 95: 65–79.

Pimentel, D. 1963. Introducing parasites and predators to control native pests. *Can. Ent.* 95: 785–792.

Regev, U., Gutierrez, A. P. and Feder, G. 1976. Pests as a common property resource: a case study of alfalfa weevil control. *Amer. J. Agric. Econ.* 58: 186–197.

Renvoize, B. S. 1973. The area of origin of *Manihot esculenta* as a crop plant—a review of the evidence. *Econ. Bot.* 26: 352–360.

Schulthess, F., Bäumgartner, J. U., Delucchi, V. and Gutierrez, A. P. 1991. The influence of the cassava mealybug, *Phenacoccus manihoti* Mat.-Ferr. (Hom., Pseudococcidae) on yield formation of cassava, *Manihot esculenta* Crantz. *J. Appl. Ent.* 111: 155–165.

Séminaire-atelier International sur la lutte Biologique Contre la Cochenille Farineuse des Arbres Fruitiers *Rastrococcus invadens*. 1987. Lomé: Protection des Végétaux. Unpubl. report.

Singh, L. B. 1968. *The Mango. Botany, Cultivation and Utilization.* World Crop Books. Leonard Hill, London. 438 p.

Slobodkin, L. B. 1988. Intellectual problems of applied ecology. *BioScience* 38: 337-342.

Steiner, K. G. 1982. *Intercropping in Tropical Smallholder Agriculture with Special Reference to West Africa.* GTZ, Eschborn. 303 p.

Stern, V. M., Smith, R. F., van den Bosch, R. and Hagen, K. S. 1959. The integration of chemical and biological control of the spotted alfalfa aphid. The integrated control concept. *Hilgardia* 29: 81–101.

Tamô, M. 1991. "The interactions between cowpea (*Vigna unguiculata* Walp.) and the bean flower thrips (*Megalurothrips sjostedti* Trybom) in Rep. of Benin. Ph.D. thesis. Switzerland, ETH Zrich. 136 p.

van Alphen, J. J. M., Neuenschwander, P., van Dijken, M., Hammond, W. N. O. and Herren, H. R. 1989. Insect invasions: the case of the cassava mealybug and its natural enemies evaluated. *The Entomologist* 108: 38–55.

van Alphen, J. J. M. and Vet, L. E. M. 1986. "An evolutionary approach to host finding and selection." in *Insect Parasitoids.* J. Waage and D. Greathead (eds.). Academic Press, London. pp. 23–61.

van den Bosch, R. 1978. *The Pesticide Conspiracy.* Doubleday, New York. 223 p.

van Dijken, M., Neuenschwander, P., van Alphen, J. J. M. and Hammond, W. N. O. 1991. Sex ratios in field populations of *Epidinocarsis lopezi*, an exotic parasitoid of the cassava mealybug , in Africa. *Ecol. Ent.* 16: 233–240.

van Emden, H. F. 1991. The role of host plant resistance in insect pest mismanagement. *Bull. Ent. Res.* 81: 123–126.

van Lenteren, J. C. 1980. Evaluation of control capabilities of natural enemies. Does art have to become science? *Neth. J. Zool.* 30: 369–381.

Vinson, S. B. and Barbosa, P. 1987. "Interrelationships of nutritional ecology of parasitoids," in *Nutritional Ecology of Insects, Mites, Spiders, and Related Invertebrates.* F. Slansky and J. G. Rodriguez (eds.). John Wiley and Sons, New York. pp. 673–695

Waage, J. K. 1986. "Family Planning in Parasitoids: Adaptive Patterns of Progeny and Sex Allocation," in *Insect Parasitoids.* J. Waage and D. Greathead (eds.). Academic Press, London. pp. 63–95.

Walker, B. H. and Norton, G. A. 1982. Applied ecology: Towards a positive approach. II. Applied analysis. *J. Envir. Manage.* 14: 325–342.

Walker, P. T., Heydon, D. L. and Guthrie, E. J. 1985. "Report of a survey of cassava yield losses caused by mealybug and green mite in Africa, with special reference to Ghana." Tropical Development and Research Institute, London. 83 p.

Weil, R. R. 1990. Defining and using the concept of sustainable agriculture. *J. Agron. Educ.* 19: 126–130.

Weiss, A. and Robb, J. G. 1989. Challenge for the future: Incorporating systems into the agricultural infrastructure. *J. Prod. Agric.* 2: 287–289.

Wilbert, H. 1980. Der Einfluss resistenter Pflanzen auf die Populationsdynamik von Schadinsekten. *Z. Ang. Ent.* 89: 298–314.

Williams, D. J. 1986. *Rastrococcus invadens* sp. n. (Hemiptera: Pseudococcidae) introduced from the Oriental Region to West Africa and causing damage to mango, citrus and other trees. *Bull. Ent. Res.* 76: 695–699.

Willink, E. and Moore, D. 1988. Aspects of the biology of *Rastrococcus invadens* Williams (Hemiptera: Pseudococcidae), a pest of fruit crops in West Africa, and one of its primary parasitoids, *Gyranusoidea tebygi* Noyes (Hymenoptera: Encyrtidae). *Bull. Ent. Res.* 78: 709–715.

Wilson, E. O. and Peter, F. M. (eds.). 1988. *Biodiversity.* National Academy Press, Washington. 521 p.

Wodageneh, A. 1989. "Constraints confronting national biological control programs," in *Biological Control: A Sustainable Solution to Crop Pest Problems in Africa.* J. S. Yaninek and H. R. Herren (eds.). IITA, Ibadan. pp. 166–172.

Yaninek, J. S., Mégevand, B., de Moraes, G. J., Bakker, F., Braun, A. and Herren, H. R. 1992. Establishment of the neotropical predator *Amblyseius idaeus* (Acari: Phytoseiidae) in Benin, West Africa. *Biocontrol Sci. Technol.* (in press).

Zwölfer, H., Ghani, M. A. and Rao, V. P. 1976. "Foreign exploration and importation of natural enemies," in *Theory and Practice of Biological Control.* C. B. Huffaker and P. S. Messenger (eds.). Academic Press, New York. pp. 189–207.

About the Book and Editor

Top-down approaches to pest management, relying on agrochemical inputs that can be scarce, expensive, ecologically toxic, or inaccessible, have repeatedly failed to solve pest problems that affect small farmers in developing countries. *Crop Protection Strategies for Subsistence Farmers* offers an alternative. Drawing on examples from Latin America, Africa, and Southeast Asia, this volume describes strategies that rely on farmers' knowledge and participation, local resources, and alternative low-input methods as a sensitive approach to developing and implementing pest management schemes adjusted to farmers' needs and their socioeconomic and agroecological conditions.

The chapters explore knowledge systems that farmers apply to pest problems, describe traditional pest management techniques, and discuss farmers' perceptions about pests. In addition, several contributors describe methodologies on how to diagnose pest problems quickly and to design simple but effective pest control methods that rely on biological as well as cultural management techniques. An analysis of the cultural, socioeconomic, and environmental advantages of alternative methods is provided. Some of the difficulties and challenges encountered by researchers in the development of bottom-up IPM strategies are also presented.

Miguel A. Altieri is associate professor of agroecology in the Division of Biological Control in the College of Natural Resources at the University of California–Berkeley.

About the Contributors

Candida B. Adalla is an assistant professor at the Department of Entomology, University of the Philippines at Los Banos. She is project leader of an IDRC-funded "Integrated Nutrient and Pest Management, Extension and Women" Project, which is working closely with farm families on the realities of implementing INPM. She is also active in research on alternatives to chemical pest control, particularly on breeding for insect resistance in rice, legumes, and cotton crops. She has also been involved in research and regional networking on issues of Women in Development.

Miguel A. Altieri is an associate professor and entomologist at the Division of Biological Control, University of California, Berkeley. He is also research and training adviser for the Latin American Consortium on Agroecology and Development (CLADES) based in Santiago, Chile. He teaches courses in agricultural ecology, rural development, and integrated pest management in developing countries. He also conducts international short courses in the United States and Latin America on biological control and agroecology in addition to researching methods to enhance biological control agents of pests in annual agricultural systems and orchards, and ways of improving productivity of traditional farming systems in the Third World.

A. C. Bellotti is an entomologist with the Cassava Program of the Centro Internacional de Agricultura Tropical (CIAT), Cali, Colombia. He received his B.S. degree from New Mexico State University and his Ph.D. degree from Cornell University. He worked with Baxter Laboratories for four years, the United States Peace Corps for five years and has been a member of the CIAT Scientific Staff since 1974. His research efforts have concentrated on the biology and ecology of arthropod pests, biological control, and host plant resistance.

A. R. Braun is an IPM specialist with the Cassava Program of the Centro Internacional de Agricultura Tropical (CIAT), Cali, Colombia. She received her B.S. degree in Plant Science from Cornell University in 1979 and her Ph.D. in Ecology from the University of California, Davis in 1986. After spending 1987 at CIAT as a Fulbright Fellow, Dr. Braun joined the CIAT scientific staff in 1988. Her research has concentrated on the biology and ecology of cassava mites and their natural enemies and on integrated pest management.

Clifford S. Gold received his education at the University of California, Berkeley, receiving a B.Sc. in Conservation of Natural Resources, an M.Sc. in Entomology (insect ecology), and a Ph.D. in Entomology (biological control). His doctoral dissertation was conducted at CIAT in Colombia, researching intercropping and cassava whiteflies. Dr. Gold is currently project coordinator and entomologist of the IITA Highland Banana Project based in Kampala, Uganda. This project emphasizes diagnostic surveys to determine the incidence of weevil, nematodes, and diseases on cooking and beer bananas in Uganda and to elucidate the ecological and farm management factors that influence their distribution and importance. The long-term objectives of this project include feasibility studies on biological and cultural controls of these pests and diseases.

James A. Litsinger gruduated from the University of California, Berkeley, in Agricultural Entomology in 1964, then joined the Peace Corps and worked in a rural development program in Brazil. Later he received his M.S. and Ph.D. at Wisconsin University and rejoined the Peace Corps in Tonga South Pacific working with a research team at the central agricultural experiment station developing IPM Programs for the country's major crops. In 1974 he joined the International Rice Research Institute, where he has worked on the Farming Systems Research and Development team in pest management. He is now a private consultant in international agricultural development living in Dixon, California.

J. C. Lozano is a pathologist with the Cassava Program of the Centro Internacional de Agricultura Tropical (CIAT), Cali, Colombia. He received his B.S. degree from the National University of Colombia in Palmira and his M.S. and Ph.D. degrees from the University of Wisconsin, Madison. He worked for eight years with the Colombian Institute of Agriculture and has been a member of the CIAT scientific staff since 1971. He has worked on many aspects of cassava diseases during the past 22 years.

Kenneth T. MacKay is director general of ICLARM, an international fisheries center in Manila, Philippines. He was formerly with the Canadian aid agency, International Development Research Centre, as Coordinator of the Environment and Sustainable Development Unit. He has been involved with a number of projects and researchers in southeast Asia examining the issues of sustainable agriculture and alternative pest management. He also has a long history of involvement in these issues in Canada as a researcher and practitioner.

Peter Neuenschwander is a member of the Plant Health management Division and leader of the biological control program at the International Institute of Tropical Agriculture (IITA) in Cotonou, Republic of Benin. There he directs research on cassava mealybug, mango mealybug, and water hyacinth in Africa and coordinates biological control activities on cassava, cowpea, maize, banana, and against grasshoppers and locusts. The two programs in his division, the Host Plant Resistance Program and the Habitat Management Program, focus on assisting national programs in the development of sustainable and productive systems for subsistence farmers, thereby improving the nutritional status and well-being of low-income people in sub-Saharan Africa.

Agnes Rola, an agricultural economist, is assistant professor at the Center for Policy and Development Studies at the University of the Philippines at Los Banos. She is the study leader of the policy component of the IDRC-funded "INPM, Extension and Women" Project. She is very actively involved in the research on the issues of environmental effects of pesticides in the Philippines and is the author of *Pesticides, Health Risks and Farm Productivity: A Philippine Experience.*

Index

IAPSC. *See* Inter-African Phytosanitary Council

IARCs. *See* International Agricultural Research Centres

ICA. *See* Instituto Agropecuario Colombiano

ICIPE. *See* International Center of Insect Physiology and Ecology

IDRC. *See* International Development Research Centre

IIBC. *See* International Institute of Biological Control

IITA. *See* International Institute of Tropical Agriculture

Inayatulla, C., 75

Income, 13, 14, 15, 18, 66, 112

India
  IIBC station, 156
  pest control, 70
  rice and wheat rotation, 71, 73, 74, 75(fig.), 77(fig.)

Indonesia
  fish and rice culture, 68
  intercropping, 83, 84
  IPM, 29, 39, 68
  pest control, 4, 7, 26, 52, 83, 84, 87
  pesticide ban, 25, 33, 104
  pests, 84
  pests as food, 3
  rice disease, 58
  See also Java

Industrial starch, 120

Inoculum load, 70

Insectaries, 149, 150

Insecticides. *See* Pesticides

Insects, 2, 4, 8, 21, 58, 78, 144–145, 148
  beneficial, 13, 14, 24, 34(table), 147
  and diseases, 55–56
  sucking, 55
  and weed interaction, 53–55

Institut Français de Recherche pour le Développement et Coopération

Instituto Agronómico do Paraná (IAPAR) (Brazil), 107, 108

Instituto Agropecuario Colombiano (ICA), 119

Integrated pest management (IPM), 29, 47, 56, 85, 143
  constraints, 33–37, 39, 85–86, 104
  holistic approach, 47, 86, 89, 113
  and inappropriate technology, 1, 18, 89
  and pesticides, 29, 33, 49, 71, 104, 108, 144, 164

re-examination of, 1–2, 18, 39, 89
  success characteristics, 105, 109, 118
  technology, 8, 19, 40, 46, 86, 89, 105, 106, 108, 109, 110–111, 113
  training, 9–10, 25, 39
  unit, 46
  See also under Farming Systems Research; *individual crops and countries*

Inter-African Phytosanitary Council (IAPSC) (Cameroon), 146–147, 148, 161

Intercropping, 14, 15, 21, 143
  and pest control, 5(table), 6, 13, 84, 123, 127, 128(table), 132–133, 137, 138
  See also under Cassava

International Agricultural Research Centres (IARCs), 27

International Center of Insect Physiology and Ecology (ICIPE), 158, 161

International Development Research Centre (IDRC) (Canada), 26

International Institute of Biological Control (IIBC) (Great Britain), 28, 146, 147, 148, 149, 156, 161

International Institute of Tropical Agriculture (IITA), 147, 148, 149, 150, 153, 157, 158, 159, 160, 162, 165
  Biological Control Program, 146, 158, 161
  Plant Health Management Division, 148, 161
  training program, 161

International Rice Research Institute (IRRI), 27, 32
  experiment station, 58
  research, 59, 63

Interrow cultivation, 79

Inundative releases, 28, 106, 109, 155

IPM. *See* Integrated pest management

IR8 (rice), 55

IR50 (rice), 88

IR58 (rice), 60, 63(table)

IR74 (rice), 60, 63(table)

IRRI. *See* International Rice Research Institute

Irrigation, 10, 11
  and pests, 21, 24
  See also Rice, irrigated

Isoprothiolane (pesticide), 56

Japan
  fish and rice culture, 68, 70
  pest control, 70
  pesticide use, 68

Printed and bound by CPI Group (UK) Ltd, Croydon, CR0 4YY

23/10/2024

01778241-0016